D0395239

Reference Phenomena Ranked by Approximate Size

SIZE (in meters)	PHENOMENON (size = width or diameter unless otherwise specified)
1,000,000 m (10^6 m)	Distance between New York and Cincinnati
100,000 m (10^5 m)	Lake Michigan
10,000 m (10^4 m)	Mt. Everest (height)
1,000 m $(10^3 \text{ m} = 1 \text{ km})$	George Washington Bridge (longest span)
100 m (10^2 m)	Football field (length)
10 m (10^1 m)	Tennis court
1 m (10^0 m)	Tall man's stride (length)
0.1 m (10^{-1} m)	Man's hand
0.01 m $(10^{-2} \text{ m} = 1 \text{ cm})$	Sugar cube
0.001 m $(10^{-3} \text{ m} = 1 \text{ mm})$	Dime (thickness)
0.0001 m (10^{-4} m)	Human hair
0.00001 m (10^{-5} m)	Spore
0.000001 m $(10^{-6} \text{ m} = 1 \text{ micron})$	Soap-bubble film (thickness)
0.0000001 m (10^{-7} m)	Particle of cigarette smoke
0.00000001 m (10^{-8} m)	Poliomyelitis virus
0.000000001 m $(10^{-9} \text{ m} = 1 \text{ nm})$	Small ion
0.0000000001 m $(10^{-10} \text{ m} = 1 \text{ angstrom})$	Hydrogen atom

A Field Guide
to the
Atmosphere

THE PETERSON FIELD GUIDE SERIES®
Edited by Roger Tory Peterson

1. Birds (eastern) — *R.T. Peterson*
1A. Bird Songs (eastern) — *Cornell Laboratory of Ornithology*
2. Western Birds — *R.T. Peterson*
2A. Western Bird Songs — *Cornell Laboratory of Ornithology*
3. Shells of the Atlantic and Gulf Coasts, W. Indies — *Abbott & Morris*
4. Butterflies (eastern) — *Opler & Malikul*
5. Mammals — *Burt & Grossenheider*
6. Pacific Coast Shells (including Hawaii) — *Morris*
7. Rocks and Minerals — *Pough*
8. Birds of Britain and Europe — *Peterson, Mountfort, & Hollom*
9. Animal Tracks — *Murie*
10. Ferns (ne. and cen. N. America) — *Cobb*
11. Eastern Trees — *Petrides*
11A. Trees and Shrubs — *Petrides*
12. Reptiles and Amphibians (e. and cen. N. America) — *Conant & Collins*
13. Birds of Texas and Adjacent States — *R.T. Peterson*
14. Rocky Mt. Wildflowers — *Craighead, Craighead, & Davis*
15. Stars and Planets — *Pasachoff & Menzel*
16. Western Reptiles and Amphibians — *Stebbins*
17. Wildflowers (ne. and n.-cen. N. America) — *R.T. Peterson & McKenney*
18. Birds of the West Indies — *Bond*
19. Insects (America north of Mexico) — *Borror & White*
20. Mexican Birds — *R.T. Peterson & Chalif*
21. Birds' Nests (east of Mississippi River) — *Harrison*
22. Pacific States Wildflowers — *Niehaus & Ripper*
23. Edible Wild Plants (e. and cen. N. America) — *L. Peterson*
24. Atlantic Seashore — *Gosner*
25. Western Birds' Nests — *Harrison*
26. Atmosphere — *Schaefer & Day*
27. Coral Reefs (Caribbean and Florida) — *Kaplan*
28. Pacific Coast Fishes — *Eschmeyer, Herald, & Hammann*
29. Beetles — *White*
30. Moths — *Covell*
31. Southwestern and Texas Wildflowers — *Niehaus, Ripper, & Savage*
32. Atlantic Coast Fishes — *Robins, Ray, & Douglass*
33. Western Butterflies — *Tilden & Smith*
34. Mushrooms — *McKnight & McKnight*
35. Hawks — *Clark & Wheeler*
36. Southeastern and Caribbean Seashores — *Kaplan*
37. Ecology of Eastern Forests — *Kricher & Morrison*
38. Birding by Ear: Eastern and Central — *Walton & Lawson*
39. Advanced Birding — *Kaufman*
40. Medicinal Plants — *Foster & Duke*
41. Birding by Ear: Western — *Walton & Lawson*
42. Freshwater Fishes (N. America north of Mexico) — *Page & Burr*
43. Backyard Bird Song — *Walton & Lawson*
44. Western Trees — *Petrides*
46. Venomous Animals and Poisonous Plants — *Foster & Caras*
47. More Birding by Ear: Eastern and Central — *Walton & Lawson*
48. Geology — *Roberts & Hodsdon*
49. Warblers — *Dunn & Garrett*
50. California and Pacific Northwest Forests — *Kricher & Morrison*
51. Rocky Mountain and Southwest Forests — *Kricher & Morrison*

THE PETERSON FIELD GUIDE SERIES®

A Field Guide to the

Atmosphere

Text and Photographs by

VINCENT J. SCHAEFER

Director Emeritus
Atmospheric Sciences Research Center
State University of New York

and

JOHN A. DAY

Professor Emeritus
Linfield College, Oregon

Drawings by Christy E. Day

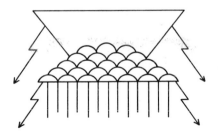

*Sponsored by the National Audubon Society,
the National Wildlife Federation,*

ar̲...

HO...

For information about this and other Houghton Mifflin trade and reference books and multimedia products, visit The Bookstore at Houghton Mifflin on the World Wide Web at http://www.hmco.com/trade/.

PETERSON FIELD GUIDES and
PETERSON FIELD GUIDE SERIES
are registered trademarks of Houghton Mifflin Company.

Library of Congress Cataloging in Publication Data

Schaefer, Vincent J.
A field guide to the atmosphere.

(The Peterson field guide series; 26)
Bibliography: p.
Includes index.
1. Atmosphere.   2. Meteorology.   3. Weather.
I. Day, John A., joint author.   II. Title.
QC863.S346   551.5   80-25473
ISBN 0-395-24080-8
ISBN 0-395-97631-6 (pbk.)

Printed in the United States of America

VB   17   16   15   14

Dedication

To Olaus Magnus, Mikhail Lomonosov, John Aitken, Benjamin Franklin, Luke Howard, Wilson Bentley, Charles Brooks, Anne Lindbergh, Irving Langmuir, Rachel Carson, Uchichiro Nakaya, Eric Sloane, M. Minnaert, and the host of others who watched the atmosphere with a continuing curiosity, a sense of wonder, and a lifelong desire to understand what they saw.

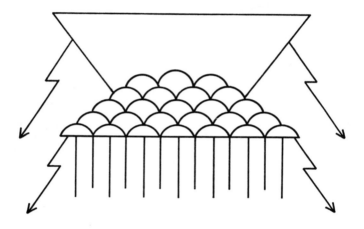

Symbol of a thunderstorm used by the Navajo of northern Arizona and western New Mexico. The heap clouds, anvil top of ice crystals, precipitation, and lightning form an elegant portrayal of this spectacular type of storm, common in the summer and essential for the survival of the inhabitants of this semiarid country.

Editor's Note

If there were not so many people interested in birds, other wildlife, and plants, we would not have the environmental movement as we know it today. The Field Guide Series, now in its fifth decade, has played no small part in laying the groundwork for this philosophical revolution, this new ecological awareness.

The *Field Guides* offer a visual approach to nature study. Although they make a bow to formal systematics by using its accepted nomenclature, they keep scientific jargon and terminology to a minimum. They rely strongly on their illustrations, which point out the distinctive features that meet the eye in the field rather than those that are obvious only under the handglass or in the laboratory.

This book, the 26th in the series, deals with a part of the environment that affects every one of us—as well as every bird, mammal, fish, insect, and plant—daily. The atmosphere is that dynamic sea of air that bathes all living things and affects their survival. We see its ever-changing cloud forms, its kaleidoscopic sky coloration, its evanescent rainbows, mists, and fogs. We feel its winds, its rains, and its snows; its heat and its cold. Yet how many of us can interpret its moods and patterns, understand clearly what they mean, or make weather predictions on the basis of them?

Dr. Vincent Schaefer and Dr. John Day, who have never lost their sense of wonder and curiosity about atmospheric phenomena, have spent a large part of their lives observing the drama of the sky. They have recorded, measured, and photographed clouds and other meteorological manifestations, not only from ground level but also from aloft, from aircraft. Their purpose in this *Field Guide* is to instill a greater awareness, so that we will think more about what goes on and take the moods of the weather less for granted.

The birder who learns to read a weather map will know when to expect a flight of raptors at Hawk Mountain or at Cape May. The atmospheric pressure and wind flow will tell him. In much the same way he can decide whether Point Pelee on Lake Erie or one of the other migration points is a good bet for warblers on a particular weekend in May or September. And, of course, if a tropical storm or a nor'easter moves in, he knows that rare seabirds are always a possibility. In a similar fashion the lepidopterist can guess with fair accuracy when certain butterflies and moths will emerge. Indeed, weather patterns affect the fortunes of many animals and plants, at times contributing to their successful reproduction, at other times condemning them to failure.

As a jogger trotting over the same country road week after week I have become especially aware of the quality of light, which can vary from day to day and even from hour to hour. As an artist and photographer I find the changing patterns of clouds very sensual. Such aesthetic experiences are in no way diminished by understanding the whys and the hows, the causes and the effects.

Quite aside from all the practical as well as the peripheral advantages of knowing something about the atmosphere, the nature-oriented person will find the recognition and interpretation of its signs and symbols fascinating. So study this *Field Guide* at home. And when you travel, by all means take it with you in your backpack, in your briefcase, or in the glove compartment of your car.

Roger Tory Peterson

Acknowledgments

The contents of this *Field Guide* are based primarily on laboratory, field, and flight experiences of the authors, and resulted from many years of observation, measurements, and photographic recording. While most of the illustrations are from our personal collections, a few have been supplied by friends and organizations as listed.

It is not possible to name all the persons and organizations who have assisted the authors in various ways over their many years of interest in the atmosphere. Dr. Schaefer, the senior author, would like to single out for special mention his parents, Peter and Rose Schaefer, who did so much to guide him toward an appreciation of the atmosphere, and also Arthur Parker, Willis Whitney, Irving Langmuir, Charles Brooks, Ted Rich, Eugene Bollay, Vernon Crudge, Carl Rossby, Harry Gisborne, Eric Sloane, Austin Hogan, Volker Mohnen, and so many others who singly and collectively played a role in his education. Much of his field experience was obtained while carrying out research projects for the General Electric Research Laboratory, the Munitalp Foundation, and the Atmospheric Sciences Research Center of the State University of New York. Dr. Day, the junior author, wishes to express gratitude to the many people along the path who have been his teachers: friends, relatives, colleagues, professors, administrators. Most of them have been fellow associates at these institutions: Colorado College, Boeing School of Aeronautics, Pan American World Airways, Oregon State University, University of Redlands, Linfield College in Oregon, and the Imperial College of Science and Technology in London. In particular he wishes to acknowledge the contribution of daughter, Christy, who did many of the illustrations. They lend a fresh touch to the appearance of the guide.

The typing of text and legends is the work of Mildred Pankrast, Lois Schaefer, Mary Lou Butchart, Ginger Bruchok, and Katherine Erickson, to whom the authors are deeply grateful.

Many of our illustrations were processed for us by Burns Photography of Schenectady and DAYPHOTO of Cheyenne, Wyoming. We also thank the U.S. Army Engineer Topographic Laboratories for permission to use some of their "Weather Extremes" records.

Much guidance, encouragement, and help in preparing this

guide was provided by Roger Peterson, Paul Brooks, Helen Phillips, James Thompson, Lisa Gray Fisher, Morton Baker, Stephen Pekich, Katharine Bernard, and Carol Goldenberg of the Houghton Mifflin staff during the nearly 20 years that have passed between our first discussions and the delivery of the finished manuscript.

Finally, to our wives, Lois Schaefer and Mary Day, for their patience and support of our efforts, we pay tribute.

For everyone's help, we are deeply grateful.

Contents

Illustrations

Credits

Eric Sloane: Rear Endpaper
Austin Hogan: Figures 5, 6
Jeanne C. Neff: Color Plate 5
Alan R. Pearson: Color Plate 17
Robert H. Day, Dayphoto Midwest: Color Plate 20
Roger Cheng: Plates 3, 4, 7
National Oceanographic and Atmospheric Administration
   (NOAA): Plates 8–25
Atomic Energy Commission: Plates 175, 176
U.S. Air Weather Service: Plate 219
General Electric Research Laboratory: Plate 238
Thomas Henderson: Plate 271
G. J. P. Barger, M.D.: Plate 277
Clyde Smith: Plate 334

This petroglyph from the early Olmec site of Chalcacingo, Morelos, Mexico, portrays an Olmec personage seated in a rock shelter from which mist is issuing. Maize, the sun, three raining cumulus clouds, thirteen large raindrops, and five objects showing concentric growth similar to hailstones are also depicted. These were carved about 1000 B.C.

Drawing by Frances Pratt, from *Chalcacingo* by Carlo T. E. Gay. Courtesy of Akademische Druck- und Verlagsanstalt, Graz, Austria.

Introduction

This *Field Guide* is written to help the curious amateur understand the various things taking place in the lower depths of the
atmospheric "sea" surrounding the earth. We have written for the
person without formal training in the atmospheric sciences and
have avoided use of mathematical formulae and complicated explanations that would have made for more precision in our explanations. Those who seek more complete explanations are encour-
aged to move on from the *Field Guide* to one of the many excellent
texts in meteorology.

We have assembled a wide selection of photographs, mostly
from our personal collections, of atmospheric phenomena, such
as clouds, optical phenomena (rainbow, etc.), and precipitation
forms. These will be of assistance to the reader in recognizing what
he sees. In addition to views of clouds from the ground, we have
included views from aircraft and satellite views of weather systems. In the center of the book is a selection of colored photographs that illustrate some of the beauty a sky watcher sees.

In addition to material on clouds, other chapters contain material on storms, small pollution particles, water in its various forms,
and weather modification. A selection of elegant but simple experiments permits the user to replicate in the home or laboratory what
is happening in the atmosphere. Once seen and understood, these
provide a direct path to the acquisition of further knowledge about
the atmosphere.

The several appendixes contain a great deal of handy tabular
and graphical information dealing with rare phenomena, water
records, and safety suggestions.

How to Use This Book

You will find *A Field Guide to the Atmosphere* helpful to your understanding of this large and varied subject. In addition to general observations about how the atmosphere is structured and how it behaves, there are many precise measurements of such things as wind speed, water concentration, temperature, particle size, and other measurable factors. In addition to simple explanations of visible phenomena — rainbows, the blue sky, lunar haloes, lightning — we have descriptions of the many changes that take place around microscopic particles. Throughout, you will find illustrations that show the effects described, and these range from color photographs of rare optical phenomena to sketches that demonstrate experiments you can try at home with simple materials. These many different kinds of information should prove interesting in themselves, and taken together may perhaps give you a new picture of our familiar — but really unfamiliar — atmosphere, one that is clearer and at the same time suggests its fascinating complexity.

The very term *Field Guide* connotes a book that can be taken out into the country, where an unfamiliar flower, animal track, or bird waits to be identified. *A Field Guide to the Atmosphere* can, of course, be used to very good advantage in the country, but atmospheric phenomena may also be observed where you live, work, or study — wherever you are. There are other differences. Birds or mammals, for example, are generally of the same size scale; species will not vary in size, one from another, much more than 10 or perhaps 100 times. As you can see from the front endpaper, this is not so with the phenomena of the atmosphere; a storm system — which is too large to be observed in toto except by satellite — is a million million times larger than a raindrop. Less extreme examples abound.

In addition, the flower or bird does not change rapidly with time (even though the bird is mobile). The atmosphere, however, is a dynamic system and is in constant change; clouds, the most commonly observable effect, change from moment to moment with never a repetition of form.

Organization With these facts in mind, we have tried to organize the information in this *Field Guide* for maximum usefulness. Although some subjects are very closely interrelated (clouds, particles, and water, for instance) and will therefore be mentioned together quite often, the chapter titles listed in *Contents* will give a clear idea of the basic breakdown of information presented, and will suggest something of our approach to explaining the various types of atmospheric phenomena, how they interact, and how they can best be observed. In addition to descriptions of how the ingredients of the atmosphere work together, there are schematic diagrams — such as in the endpapers — that will help to orient you by showing the size relationships of the things described and that also show such items as typical cloud profiles in idealized form, the simple experiments, and other things that can be better understood with an illustration. Among the most important things illustrated in this way are the enormously varied bodies and forces at work: particles, water, chemical reactions, and light energy in its many forms, to name a few.

Illustrations You will find the photographs helpful as well; not only the color and black-and-white photographs of clouds, rainbows, false suns, and other phenomena visible to the naked eye, but also enlargements and photomicrographs of such items as hailstones, dust particles, and ice crystals. Dates are included in some photo captions in case a curious reader wishes to go to the meteorological records and correlate the photo with the meteorological situation of the day.

Of special interest to everyone who has enjoyed flying and has noted the different aspects of clouds in this environment, a chapter and many photographs are devoted to clouds as they are seen from airplanes.

Simple Experiments A chapter describing simple experiments gives you the opportunity to try some of the effects described with the use of fairly simple materials that are easily acquired. Many of the devices and procedures are very simple and are very rewarding. Pointers for the use of camera equipment for capturing atmospheric effects are included, and should be very useful for the photography enthusiast.

Glossary A glossary composed of various terms that have special application to the atmosphere is included in this guide; many of these terms are also explained in the text as they are encountered.

Appendixes These provide a great deal of valuable information — especially tabular material and figures — that tell, for example, what the wind-chill factor will be under any given conditions. Other appendixes explain certain complex phenomena mentioned in the text, give international weather symbols and temperature records, and provide a handy guide for making conversions to the metric system and vice versa. Two of the appendixes give extremely valuable safety information about tornadoes and lightning.

Measurements The United States is the principal country in which the foot/pound/Fahrenheit system of measurement is still in use; even the British — from whom we inherited the system — are further down the road toward complete conversion to metric measure. The U.S. is committed to change, however, and the metric system is already widely used throughout the scientific establishment, with the change also well underway in elementary education and in certain industries. TV weather roundups now carry temperatures in both Fahrenheit and Celsius degrees, and some road signs show distances in kilometers. In some service stations gasoline is being pumped in liters.

In this *Field Guide* metric measurements are the primary ones used. They have been converted to U.S. units when the authors considered it useful and appropriate. Conversion tables appear in Appendix 4.

In most instances we have used the unit *microns* rather than the newer term *micro-meters* since the former is a simpler term and is in universal use in the atmospheric sciences for sizing very small objects like cloud droplets and condensation nuclei. Some diagrams use micro-meters (μm) to conform to international practice.

In general we use the metric form for barometric pressure, i.e., mm of mercury (Hg), even though the daily weather report on radio or TV usually expresses barometric pressure in inches of mercury. 760 mm Hg is the same as 29.92 in. Hg. Both figures are expressions of standard sea-level atmospheric pressure. Occasionally the unit millibar (mb) will be used.

Following are some specific suggestions about things a reader can do to reap the greatest rewards from the *Field Guide to the Atmosphere*.

Use the *Field Guide* with Daily Weather Maps By clipping the daily weather maps published in the local paper and marking them with the plate numbers of clouds that appeared during the day, the *Field Guide* user will soon discover a correlation between the frontal systems, high and low pressure centers, winds and precipitation patterns, and the clouds that are illustrated. This will help in forecasting the next day's weather.

Appreciate the Range-of-Size Scale By all means take a detailed look at the size range of phenomena. The relative sizes of various atmospheric phenomena are shown on the front endpaper. Note that the atmosphere, although global in breadth, is an extremely shallow film of air. The lowest layer — the troposphere — is only a few kilometers thick, ranging from 15 km at the equator to 10 km at the poles. This is only ¼ of 1% of the earth's radius, yet all terrestrial life is sustained within this paper-thin layer.

Become a Weather Diarist We hope this book will encourage you to take up the practice of keeping a weather diary — a daily log of the observed weather. Keeping a diary will help you make systematic observations of selected aspects of the weather — starting, for example, with types of clouds and amounts of rainfall. This practice might be expanded to include readings of temperature, pressure, humidity, visibility, and wind direction and speed.

Observing the weather regularly can become such an ingrained habit that the day is incomplete without having made the appropriate observations. A good friend of Dr. Day's has become the local weather seer in the town of McMinnville, Oregon, because he has kept a personal log of rainfall and temperature for the past 30 years. Each week the local newspaper carries the summary of daily precipitation from his rain gauges.

Some weather diarists have standardized their instruments and have become cooperative observers for the National Weather Service. They make their observations on the same strict time schedule as the professional observers, and these observations become a part of the official climatological record.

Note Changes in Visibility Visibility is an important indicator of how settled the atmosphere is: that is, poor visibility indicates that the local atmosphere is relatively settled and stable, while good visibility indicates the converse. Lowered visibility is the result of an increasing load of small pollution particles that intercept and scatter light that comes from the object being viewed. You can identify certain landmarks such as mountains, hills, and prominent buildings, and determine their distances from your home. The degree to which they can be seen establishes the visibility; on one day it may be more than 20 km (12½ mi.), and on another it may be reduced to a kilometer or less. This is an appropriate item to enter in your weather log.

Study the Sky Many people do not see the sky, even when they look at it. In the past, sailors and farmers studied the sky early and often; their livelihood depended on their being weatherwise, and many developed great skill as 24-hour forecasters despite the primitive state of the science of meteorology. Anyone today can study the sky via the satellite pictures of clouds displayed regularly on the TV weather report. Combining this kind of once-removed observation with actual consistent and careful firsthand observations of the sky is the way to develop your own skill in predicting the weather on a short-term basis. Even the most knowledgeable forecasters make mistakes in judgment, so don't be discouraged if it rains when you forecast clear, sunny weather.

Identify the Clouds The many photographs in this *Field Guide* will help with cloud identification. Compare them with the clouds that you observe. The photographs chosen for the book are, for the most part, of "pure" types — in learning cloud forms that is the place to start. The sky itself frequently contains cloud mixtures that make identification an exercise in frustration, even for the experts; the amateur observer can take solace in this. We can only recommend, "Keep trying — there is always another simpler sky." Practice and increasing familiarity will make it easier to judge cloud height (usually difficult for the beginner) and separate one cloud group from another.

Identify Clouds from the Air When you fly, sit by a window and have a "box seat" for the atmosphere. At 11 km (35,000 ft.) you can look down on most clouds, or look directly at a towering cumulonimbus as the airplane skirts its perimeter. It would be desirable for all airlines to have this *Field Guide* as standard equipment for their passengers. However, until this happens you may want to consider taking along your own copy so that you can refer to Chapter 5, *Clouds from Airplanes.*

Investigate What Falls from the Sky Dr. Schaefer has spent a lifetime investigating the small particles that are suspended in the air and the so-called hydrometeors — liquid or solid particles — that fall from the sky. This kind of observation can be done at all levels of sophistication. The sizes of raindrops coming from unstable heap clouds may be compared with those from stable layer clouds. Hailstone diameters may be measured with simple calipers. A hailstone can be sawed into thin slabs in a deep-freeze to study its layered structures and infer information about the cloud from which it has fallen. Photography of snow crystals is a bit more complicated, but is well within the grasp of the interested amateur. (See the replica method described in Chapter 10.)

Do Simple Experiments Dr. Schaefer has selected a few simple experiments from his large collection for inclusion in the guide. These will be valuable to users who wish to make some investigations on their own, and particularly to teachers at all levels from grade school to university who are looking for student projects. There is no better way to understand the complex relationships that exist between particles, moisture, temperature, air motion, the formation of clouds, the properties of an inversion, and the effects of pollution on cloud stability.

Familiarize Yourself with Weather Safety Information It may save your life, or the life of another, if you are knowledgeable about weather safety recommendations. In particular, you are referred to Appendixes 17, 19, and 20 dealing with Wind Chill and Hypothermia, Tornadoes, and Lightning. Learn the warning signals of these 3 phenomena, which have potentially fatal consequences, and be prepared to act wisely.

Carry a Copy of the *Field Guide* When planning a trip of any kind, consider taking this *Field Guide.* The dynamic atmosphere is always with you and there may be something happening in it *today* that you will wish to check in your guide.

A Field Guide
to the
Atmosphere

1

The Global Atmosphere

The global atmosphere covering the earth has been likened to the skin of an apple. In this thin layer are contained the life-sustaining oxygen (20.95%) required by humans and other creatures; the carbon dioxide (0.03%) so essential for plant growth; the nitrogen (78.09%) also needed for chemical conversion to plant nutrients; the trace gases such as neon, helium, methane, krypton, hydrogen, xenon, and ozone; and varying amounts of water vapor and airborne particles.

The force of gravity holds about half the weight of the whole atmosphere — a fairly uniform mixture of these gases — in the lowest 5.49 km (18,000 ft.); 98% of the material in the atmosphere is below 30.5 km (100,000 ft.). Putting the matter in a slightly different way, the atmospheric pressure drops from 1000 millibars at sea level to 10 mb at 30.5 km. From 30.5 km to 61 km (100,000–200,000 ft.) the pressure drops from 9.9 mb to 0.1 mb and so on.

Although the atmosphere is considered to have a thickness of 64–80 km (40–50 miles) we are primarily concerned in this *Field Guide* with the troposphere, the part of the earth's atmosphere that extends from the surface to a height of about 8 km (26,200 ft.) above the poles, about 11 km (36,000 ft.) in mid latitudes, and 16 km (52,500 ft.) over the equator. Above the tropopause (the top of the troposphere) is the stratosphere, a region that increases in temperature with altitude until it reaches its upper limit of 80 km (260,000 ft.).

In the stratosphere, ozone is produced by the intense ultraviolet radiation from the sun; it is the absorption of the sun's radiation by ozone that warms the stratosphere and makes it quite stable — it is also this absorption by ozone that protects living things on earth from destructive ultraviolet rays. The total amount of ozone in the atmosphere is small: if compressed to a liquid layer over the globe at sea level, it would have a thickness of less than 5 mm (³⁄₁₆ in.); in the troposphere the average concentration of this protective layer is about 20 parts of ozone per billion parts of air. Occasionally this concentration is 5–10 times higher than the average — this occurs when air from the stratosphere is carried toward the earth by the proximity of a jet stream.

Water Vapor

The most variable as well as the most remarkable of the atmospheric trace gases is water vapor. Much of this book describes the ways in which water vapor, in its various forms, controls the behavior of the air surrounding our planet. Unlike the other trace gases, water exists in the atmosphere in gas, liquid, and solid form, and adds and extracts heat from the air whenever it changes from one to another.

Water vapor and airborne particles are essential for the stability of the global ecosystem. Their variations and interactions, combined with the global circulation of the atmosphere, produce the world's weather (including its clouds and precipitation) and are responsible for the blue-green-white appearance of the earth as seen from space.

How We Perceive the Atmosphere

The natural atmosphere has many features that everyone knows and enjoys: the blue of the sky, the clarity of the air, the brilliance of the stars at night, the color of the rainbow, the smell of the sea, the fresh air following a thunderstorm, the red of the sunset, and the symmetry of a snow crystal. In the same category as the deep blue of the clear sky is the blue color of glacial ice and the soft turquoise of the shadows in holes in freshly fallen snow.

But the atmosphere has another face. The terrible destructiveness of a tornado, a hailstorm, or a hurricane, the wearying monotony of the persistent gusty Foehn or Mistral winds, the roar of a wildfire sweeping up a mountain slope — these are some of the terrifying examples of the disturbed atmosphere.

Cities in general are poor places to survey the most attractive features of the atmosphere. Apart from the pollution that rises from the city to mask the stars and make the visible sky a dirty brown or at best a pale blue, the buildings of the city obstruct the view. Except for the panorama that can be enjoyed from the top of a skyscraper, the observable atmosphere is not very available to the city dweller. This *Field Guide* is intended to encourage the reader to seek the natural surroundings of the countryside; by doing so, the reader will be better able to understand the beliefs and intentions of those who would preserve the environment through intelligent land-use practices and control of all forms of pollution.

Fortunately for the survival of all earth's organisms, the atmosphere has many self-healing properties. Clouds are its air-cleaning agents, and the remarkable global circulation systems constantly purge the air of its foreign matter; it is only when air contamination overloads the system that living organisms are faced with serious trouble.

The Range in Size of Atmospheric Components

One of the intriguing aspects of the atmosphere is the way its components vary in size; at one end of the scale are molecules, and at the other are gigantic storm systems and the entire atmosphere itself, so that sizes may range from 0.0001 micron to 10,000 km. This range of 17 orders of magnitude may be more easily imagined if you refer to the front endpaper (very conveniently, human beings happen to be about midway in size between these extremes).

Reduced Visibility in the Atmosphere

The reduction in visibility of distant objects is caused by a variety of factors and often the combination of several. The most effective particle shape and size for scattering visible light is a spherical particle with a diameter of 0.6 micron. These particles scatter light effectively in nearly all directions, but the maximum scattering occurs on a line between the observer and the light source. This accounts for our being able to look directly at the sun without hurting our eyes under the right conditions — when the air contains enough particles to restrict visibility significantly, and the sun is 10° or less above the horizon. In clean air, as soon as the sun rises it becomes extremely hazardous to view directly.

Types of Airborne Particles

The atmosphere contains a wide variety of airborne objects and substances. The largest of these range in size from the debris of tornadoes, waterspouts, volcanoes, burning embers from forest fires, and tumbleweeds, to spider webs, seed parachutes, soil particles, pollen grains, and other living microscopic forms.

The smaller light-scattering particles include fragments of rock made airborne by dust devils and gusty winds, salt and spray from the breaking of bubbles at sea, forest-fire smoke, and particles that produce the blue haze often seen over forested mountains. The greatest number of airborne particles are so small as to be invisible. These submicroscopic particles are formed by the condensation of vapors, the chemical combination of reacting molecules, the photochemical effects produced by the ultraviolet radiation from the sun, and the electrical and other ionizing forces that come from thunderstorms, cosmic rays, and radioactivity.

The Formation of Airborne Particles

Airborne particles are produced by 2 very different mechanisms. The larger particles are fragments of still larger ones which, through weathering, mechanical breakage, solution, or some other attrition process, finally become small enough to float in the at-

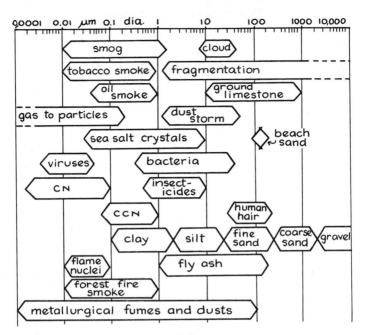

Fig. 1 Size range of airborne particles.

mosphere when made airborne by the wind. These particles may sometimes be reduced in size to 1 micron or smaller, but when found in the air are generally 5 microns or larger, the smaller ones joined to the larger ones by adhesion, electrical forces, capillarity, or surface tension. Many of these larger particles are of natural origin, although some are manmade, resulting from processes like the production of cement, quarrying, and strip mining.

The other major mechanism of small-particle formation begins in the vapor phase and proceeds by condensation, crystallization, or related mechanisms. The cooling of hot saturated or supersaturated vapor, the combination of chemicals, photolytic reactions, and other condensation processes typically produce such particles. Their sizes range from molecular clusters of 0.001 to 0.003 micron to about 1 micron. Under special conditions of high temperature and a rich source of condensable material, as in a volcano or the exhaust vapors in the furnace of a coal-burning power plant, some particles can grow to sizes considerably larger than 1 micron. Such conditions are rare, however, so that the common ranges occur and overlap as shown in Fig. 1. Only particles that remain airborne for periods of at least an hour tend to be of importance in atmospheric reactions. This limits the size to those smaller than about 50 microns (0.002 in.). These particles have a diameter slightly smaller than the cross section of a human hair.

Since the oceans of the world cover 70% of its surface, they are an important source of atmospheric particles produced when spray evaporates or bubbles of entrapped air burst.

The concentration of sea salt particles is low, however, so that they rarely affect visibility. The exceptions occur when all of the seawater on the spray particles does not evaporate. If the relative humidity of the air is higher than 70% the tiny particles start to grow; visibility is impaired by haze, and eventually fog may form. This is common along the seashore.

Fallout of Airborne Particles

The largest airborne particles may be in the air for minutes or hours; the intermediate sizes are likely to be suspended for hours or days; and the truly fine particles may reside in the atmosphere for weeks, months, and even years. Their falling velocity is shown in Fig. 2, while Fig. 3 depicts the time required to fall 1 km (3280 ft.) in calm air under average conditions. The residence time for particles depends on size, weight, shape, injection altitude, airflow, and the cleanup mechanisms that operate in the atmosphere — the formation and evaporation of clouds and the descent of precipitation. This *Field Guide* will provide basic information about the specific concentration and composition of atmospheric particles, the precipitation they induce, and the systems that produce them, move them through the atmosphere, and dispose of them after they have formed clouds and precipitation.

There is a remarkable balance in relative particle concentration, rarely modified to any extent by natural forces. Only with such catastrophic events as the massive volcanic eruption of Krakatoa in 1883, the prolonged droughts of the 1930s, or the massive forest fires in Canada of the late 1940s, are large-scale changes in the concentration of airborne particles noticeable. The volcanic ash that Krakatoa threw into the stratosphere caused vivid sunsets for several years; the great dust clouds from the western states obscured the sun over the eastern United States, and the smoke from the widespread fires in Canada during the late forties produced a blue moon in the Northeast; the smoke extended across the Atlantic into Europe. These effects will be described later in this chapter.

Visibility and Hazes

The interested observer always has some degree of visibility in the atmosphere available for his consideration. In the "clean" parts of the world, visibility — the condition of the atmosphere that allows

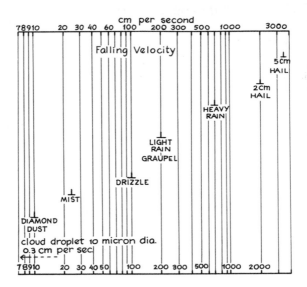

Fig. 2 Terminal falling velocity of different kinds of precipitation.

Fig. 3 Time for different kinds of precipitation to fall 1000 m (3280 ft.).

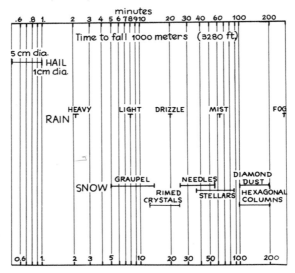

us to see the details of distant objects — is generally of excellent quality. This is the result of having a small number of visible particles floating in the air; these visible particles interfere with our vision by scattering, obscuring, refracting, or reflecting sunlight and moonlight. As the number of visible particles increases, haze forms; looking toward the horizon, we see an atmosphere that tends to be hazy in most land areas of the world. There are several varieties of haze, described in the following sections.

Blue Haze The bluish haze of the Grand Canyon, other areas of the semiarid regions of America, the Blue Mountains of Australia, and other portions of the world noted for their almost unlimited visibility, have several factors in common. These include extensive regions of vegetation like sagebrush, chaparral, evergreen, and mixed deciduous forests; the absence of forest fires, industrial activity, and other manmade particle sources, and an active interchange of air from the upper atmosphere. These hazes do not appreciably obscure visibility, but provide a beautiful opalescent blue quality to distant objects. Unfortunately, because of their liquid nature and submicroscopic size, their image has not yet been captured in the electron microscope. Present evidence indicates that many of these tiny particles are produced by a chemical combination of the naturally occurring terpenes from vegetation with the ozone carried down from the stratosphere. There are more than 11,000 terpenes known to exist — the most familiar is gum turpentine.

For most observers, the blue haze adds beauty to the landscape. It provides a 3-dimensional depth to a distant view because of the increased amount of light scattering that occurs with distance. This effect is caused by the nearly uniform scattering of light from particles just above the threshold of visibility (0.1 to 0.3 micron in diameter).

Other natural hazes of this type, such as are found in the Great Smoky Mountains of the southeastern U.S., are less blue; they have a whitish hue caused apparently by the deposition of moisture on the larger particles produced by the terpene-ozone reaction or a combination of this with the salt haze of the ocean.

Gray Haze This type of haze, which obscures distant objects by reducing their contrast with the sky, is also of natural origin. It is caused by relatively large particles (5–30 microns) of soil, salt, and other mineral material; a seasonal stimulus to this type of haze occurs in the spring, when great numbers of pollen grains and spores produce scattering effects identical in appearance to those produced by desert storms. These particles sink slowly toward the earth, but their settling velocity is often slowed by convective motions of the air and by winds.

In addition to pollen grains and spores there are other living particles ranging from bacteria to mites and larger insects. In fact, a marvelous variety of living things and their residues moves

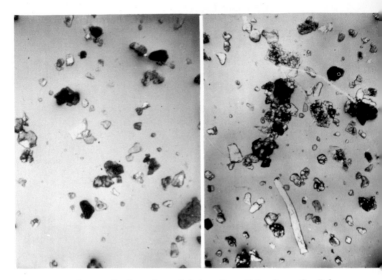

Pl.1 Natural aerosols. Sample on left is soil particles from the Sahara. The other is soil particles and natural fibers from a semiarid region of the Southwest. Fiber is 70 microns long.

through the atmosphere. Plate 1 provides photomicrographs of some typical examples.

Brown to Smoky Blue Haze This haze, which is produced by humans, is a hallmark of the industrial age. It often encompasses areas of many thousands of square miles above and downwind of cities, industrial plants, or other areas with a high density of automobile traffic and many dwellings. The multiple gas and particle sources blend to form a uniformly pervasive visibility-obscuring haze. Such hazes sometimes extend vertically for 5 km (3 mi.) or more, and under stagnant air conditions will often limit visibility to less than 10 km (6 mi.). With the passage of the Clean Air Act of 1969, a vast program to limit air pollution has been carried out by both federal and state agencies. Major emphasis has been directed at reducing the amount of noxious gases, such as the sulfur and nitrogen oxides, carbon monoxide, hydrocarbons, and airborne particles. Further discussion of this matter will be found at the end of this chapter.

In large cities such as Los Angeles and Phoenix, where the sun shines much of the time, automobile traffic tends to cause a buildup of photochemical smog, consisting of invisible particles and gases, which often has a brownish cast caused by the large amounts of nitrogen dioxide in the air. This is the end reaction of the nitrogen oxide cycle. Any hot flame, as from the internal com-

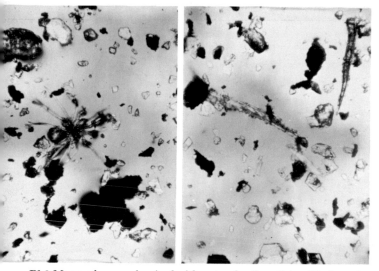

Pl.2 Manmade aerosols mixed with natural soil particles. Black particles of carbon scale are caused by burning processes ranging from autos to incinerators. Fibers 70 microns long are coated with dust.

bustion engine, furnaces, incinerators, jet engines, gas turbines, and similar sources, oxidizes the gaseous nitrogen in the air to produce these oxides. This type of haze is only encountered in association with human activities. Plate 2 shows typical aerosols of this nature.

At rare intervals volcanic dust is actually injected into the stratosphere by massive volcanic eruptions, such as at Krakatoa, Agung, and in our own country Mt. St. Helens. This stratospheric dust haze may remain at high altitudes for several years, enhancing the beauty of sunsets by scattering the sunlight for many minutes after local sunset. Unfortunately most hazes produced by man are of a different aesthetic nature. Many of the air pollution particles are larger than the blue-haze size of 0.1–0.3 micron and thus are more effective in their light-scattering capacity (the most effective light-scattering particles have a cross section of 0.6 micron).

In addition to the larger sizes, a number of the manmade particles, especially those based on the oxides of sulfur and nitrogen, are extremely hygroscopic — they have an affinity for water and other liquids. Liquid droplets are therefore often coated with an impermeable film that keeps them from evaporating, thus making the hazes highly light-absorbing and aesthetically displeasing. They are also a health hazard, since the particles and droplets are still small enough to pass through the upper respiratory system.

The Effect of Particles on Visibility

An interesting optical effect in the atmosphere, produced by the scattering of light by particles, is the crepuscular rays of the sun. These are most noticeable when the sun is low in the sky and hidden by isolated clouds. The rays appear as though produced by a giant searchlight and often fan out like the spokes of a wheel; however, since they originate from the sun, they are actually parallel with one another. The intensity of the light scattered by these rays is an indication of the concentration of fine particles suspended in the atmosphere. In areas like the North Atlantic and Central Pacific oceans, the Polar regions — wherever the air is extremely clean — crepuscular rays do not occur. In such areas the beam of a strong flashlight at night is quite invisible.

The absence of any visible haze in a mountain region can also cause problems to hikers by making relative distances very difficult to estimate. In an area without trees or other familiar height indicators, any feature of the terrain such as a cliff or hummock might be a half-mile away, or 5 miles away! In this respect, the surface of the moon must present big problems to the astronauts in estimating distances.

Particle-free air can also cause other unexpected problems. Dr. Schaefer was once in a wild cave, wading shoulder-deep in the crystalline cold waters of a pool. His companion, who had thus far remained reasonably dry, decided to light his carbide lamp to extend the life of their flashlights. When he lit a match, they were quickly enshrouded in a dense pea-soup fog. The air had been supersaturated with water vapor, and the match flame provided the nuclei that permitted this water vapor to condense to form the fog. A similar supersaturation phenomenon has been encountered in the vicinity of the hot springs and geysers at Yellowstone in the winter. The rich moisture supply, coupled with the very cold temperatures and the very clean air, are responsible; here again a flame, such as that of a match, produces a dense localized cloud.

Particles in Atmospheric Processes

The particles that participate in atmospheric processes are mostly larger than 0.1 micron, and fall into 3 categories — ice nuclei, cloud condensation nuclei, and condensation nuclei. These are particles on which water molecules will condense under the special conditions that will be described.

Ice Nuclei The largest atmospheric particles are smallest in number; they are termed ice nuclei and are typically found by microscopic examination of ice crystals. They ordinarily range from 5 to 50 microns in size, with their concentrations varying between 0.1 and 10 per liter. In unpolluted air, they are primarily

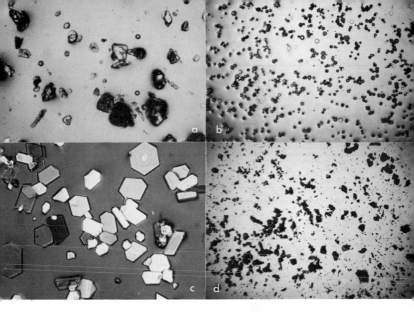

Pl.3 Typical ice-crystal embryos as are studied in the laboratory and often found in the atmosphere: a. quartz sand; b. cirrus ice crystals; c. silver iodide; d. volcanic ash.

rock or other mineral fragments, angular to rounded, and made up of volcanic ash, glacial clay, or sand. These particles, when dusted into a cold chamber (see Chapter 10) will act as nuclei for ice-crystal formation. The temperature at which they are active varies. The most effective natural substances have a threshold of ice nucleation at $-10°$ to $-12°C$, but some do not serve as ice nuclei until a water-droplet cloud is supercooled to $-20°C$ or colder. Plate 3 shows the variety of ice nuclei, and Plate 4 illustrates the types of ice crystals that grow on them.

Cloud Condensation Nuclei These are particles on which water molecules condense to form cloud droplets. Their size ranges from about 0.05 to 1 micron and concentration from 50 to 1000 or more per cubic centimeter (cc). Thus their lowest concentration is 50,000 times higher than that of ice nuclei. Their average number ranges from 20,000 to 500,000 times greater. Cloud condensation nuclei include particles of sea salt, other salty or otherwise hygroscopic particles, forest fire smoke, clays and other fine terrestrial dust particles, organic compounds produced by gaseous vapor condensation such as the terpenes (cyclic olefins from plants) reacting with ozone, as well as related products of condensation. In general, salt particles are the best cloud condensation nuclei since most of them begin to grow at relative humidities of about 70% due to their water-absorbing nature. Thus they have already started their

Pl.4 Ice crystals that have been grown on soil and ice particles in laboratory experiments: a. silver iodide; b. pentaerythritol; c. quartz sand; d. cirrus ice crystals.

Pl.5 Soil particles and salt crystals collected 1000 miles west of Africa. The salt crystals acquire water when exposed to moist air, as can be seen on the right. Largest droplet is 20 microns.

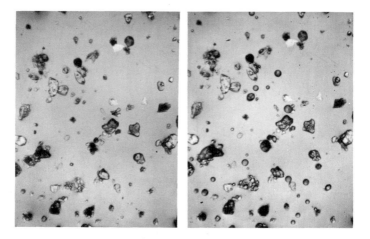

growth cycle even before reaching the normal dew-point level in the atmosphere. Plate 5 shows ocean-salt particles dry and after exposure to moist air.

Concentrations of cloud condensation nuclei range from less than 100/cc in very clean air to 1000/cc or more in air affected by man's activities or larger natural particle sources such as forest fires.

When water vapor condenses on these nuclei, the cloud droplets range in diameter from 5 to 50 microns depending on the quantity of water vapor in the air, the concentrations of active nuclei, and the amount of cooling below the dew point.

Condensation Nuclei These are submicroscopic solid or liquid particles produced by condensation from gas. Their size range extends downward from the threshold of visibility to the dimension of tiny unstable clusters of molecules: from 0.2 to about 0.002 micron. Although these particles are extremely small, they occur in relatively high numbers, and they are the origin of the cloud condensation nuclei over land areas when they collide. The ratio of cloud condensation nuclei to condensation nuclei ranges from 1:3 in extremely clean air to 1:50 in the air of a polluted city.

The cleanest air of the world is found in the central portions of the oceans, in its Polar regions, and in the stratosphere. In these areas, the total number of particles of all sizes is less than 300 per cubic centimeter. The concentration of cloud condensation nuclei ranges from less than 50 to about 100 per cubic centimeter. The ice nucleus concentrations generally range from 0.1 to 1 particle per liter. These are probably the basic concentrations of our atmosphere. Higher concentrations of fine particles of the natural atmosphere are due to the presence of active local or regional sources of gaseous vapors such as iodine, the terpenes, ammonia, and hydrogen sulfide, which in turn combine with ozone and other trace gases to produce fine particles.

When air cools to the dew point (see p. 51), the larger particles, many of which are cloud condensation nuclei, serve as centers for the deposition of water vapor. The finer particles, not active as cloud nuclei, are carried into the cloud, where many of them diffuse and deposit themselves on the cloud droplets. Once they contact the droplet, they either dissolve, remain at its surface, or sink into its volume. If the cloud droplets coalesce to form precipitation, their residue accumulates with other droplets to reach the ground with the raindrop. If the precipitation forms virga — evaporates before reaching the ground — the total residue within the precipitative element, including the original cloud condensation nucleus, remains as a much larger particle. It is then likely to fall by gravity to the earth.

If cloud droplets develop into a supercooled cloud they may freeze and grow into graupel, the spherical centers of snow crystals, or even hailstones. If, however, the supercooled droplets evaporate, their nuclei and any residues present become larger and thus may enhance the nuclei for subsequent droplet condensation. Such particles might be swept up by falling ice crystals and thus removed from the atmosphere.

The Role of Particles in the Atmosphere

The relationships just described illustrate the importance of particles in the formation of clouds in the normal atmosphere. Without particles the earth would be in serious trouble. Clouds would not form until the air became so supersaturated that spontaneous nucleation occurred. This would probably cause extremely intense cloudbursts over large areas, with resulting devastation.

The global atmosphere usually contains relatively few airborne particles. When the "normal" concentration of these particles is drastically modified by certain natural phenomena, such as extensive forest, brush, and grass fires, and massive volcanic eruptions, and much more frequently by man's activity, air pollution occurs.

Aerosol Climatology By frequent measurement of the concentration of fine particles at a specific location for a period of 5 to 10 days, the air quality in that region can be established. Between 70 and 130 measurements make a usable sample. It is desirable to make measurements 3–4 times during the year to establish seasonal variations, especially if the wind-flow patterns vary from one period of the year to another. Seasonal variations are relatively small unless the region is influenced by man's activities or by seasonal widespread fires. Fig. 4 shows the concentration of condensation nuclei found when measurements were made in a variety of locations in various parts of the world. This kind of information constitutes an "aerosol climatology" with which, over a period of years, it can be learned whether the particle levels fit into a constant pattern or are changing. This bears directly on the air quality of a region, as presented in Table 1, which shows the 7 categories established to illustrate the wide range in number that occur in the global atmosphere. Figs. 5 and 6 show the pattern of the fine particle concentrations found over the Atlantic and Pacific oceans.

The precipitation pattern on the global scale shows its relationship to the concentration and types of active cloud condensation and ice-forming nuclei in the atmosphere. Clouds over the ocean in the trade-wind regions form rain so readily that it falls long before precipitation would be expected from similar looking clouds over the land.

Cloudy air containing only a few hundred cloud condensation nuclei per cubic centimeter permits the coalescence process to proceed rapidly and efficiently in converting cloud droplets to precipitation elements large enough to fall.

At the other extreme, when air contains a great many smoke particles such as from a forest fire, water droplet clouds often form at the top of the convection column. Such clouds rarely, if ever, form precipitation, because so many cloud condensation nuclei share the available cloud water that the droplets that form are uniform and tiny, and coalescence does not occur. This is an excellent example of cloud condensation nuclei overseeding.

Measurements of the concentration of nuclei for ice crystal formation made in both the northern and southern hemispheres show that in "clean" air, unaffected by man's activity, the concentration of these particles ranges from 0.1 to 10 per liter (1000 cc). When air with a concentration of one ice nucleus per liter moves into large, growing clouds, and its droplets supercool, ice crystals that form on these nuclei are in about the right proportion to grow large enough to be affected by gravity. Thus, if 100,000 cloud droplets, each with a diameter of 12 microns, were to transfer their water molecules to a single ice crystal, that particle would be larger than 500 microns (0.5 mm) and have a terminal velocity greater than 1 meter per second.

TABLE 1

Category	Class	Particles per cc			
		0.005 μm	0.1 μm	0.5 μm	5.0 μm
1	Extremely low	0–500	50–150	0.5–2	0.01–0.1
2	Very low	501–1000	100–500	1–5	0.01–0.8
3	Low	1001–5000	201–750	1–10	0.1–2.0
4	Medium	5001–50,000	500–1200	1–10	0.2–3.0
5	High	50,001–150,000	750–3500	5–15	0.5–5.0
6	Very high	150,001–750,000	1000–8000	15–80	0.7–10.0
7	Extremely high	750,000–10 million	5000–90,000	10–25	0.5–10.0

Category	Description	Types of Places
1	Very remote places	Stratosphere, mid-oceans, Polar regions
2	Remote areas	Deserts, mountain summits, forests
3	Countryside	Farms, badlands, mountain slopes
4	Suburban	Periphery of cities, greenbelts
5	Urban	Center of cities, streets, shopping plazas
6	Highways, heavy industry	Heavy auto traffic, close to industry
7	Extreme cases, quite unnatural	Highway tunnels, closed rooms

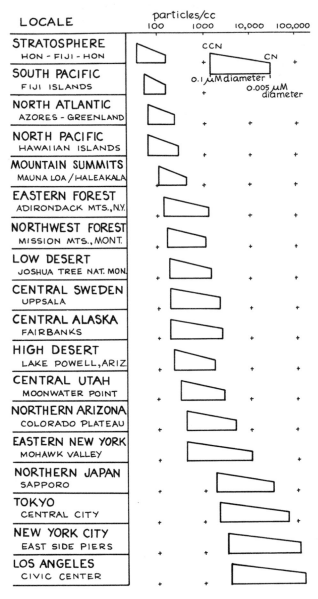

Fig. 4 Range in concentration of cloud condensation nuclei (CCN) and condensation nuclei (CN).

Manmade Effects Producing Cloud Modification

As with a superabundance of cloud condensation nuclei caused by fires or air pollution, an excessive concentration of ice nuclei will also overseed clouds.

Downwind of large urban areas in obviously polluted air, very high concentrations of ice crystals can often be seen during the winter season. The crystals are so abundant that they often produce a brilliant undersun, and are so small that they either float or have a falling velocity so low as to be hardly measurable.

The Effect of High Concentrations of Ice Nuclei In air polluted with automobile exhaust, it is not uncommon to find concentrations of potential ice nuclei and concentrations of ice crystals thousands of times higher than in clean air. Suppose that the ice nuclei are concentrated in air having 500 supercooled cloud droplets (12 microns in diameter) per cubic centimeter. The ice crystals, when they have grown at the expense of the cloud droplets, will be about 30 microns in diameter with a falling velocity of less than 1 meter per minute. Because particles of this kind tend to be very thin hexagonal plates, their falling velocity may actually be overcome by slow upward motions in the atmosphere resulting from slight warming of the air. Plate 6 shows an undersun.

Quite frequently the undersun reflection is almost round, indicating very little, if any, falling velocity since such thin plates oscillate when they are falling, producing an elongated pattern similar to the sun's reflection "path" on a slightly rippled sheet of water.

The ice nuclei from auto exhausts are probably complex lead compounds that react with atmospheric iodine to form lead iodide, a very effective ice nucleus similar to the silver iodide used in commercial cloud seeding. Other products of pollution, including cupric sulfide from smelters and iron oxide from steel mills, are potential sources of nuclei that could produce weather modification. Whether chemical factories and other industrial activities are an active source of ice nuclei is uncertain.

Problems Caused by Air Pollution These are especially difficult to anticipate because, unlike natural high concentrations of particles due to volcanism, forest fires, or dust storms, which merely provide higher concentrations of the particles normally present in the natural atmosphere, the large concentrations produced by man are unnatural and thus are likely to produce an entirely new set of chemicals, reactions, and particle-size spectra.

Thus, while some of the natural sources of gaseous emanations caused by the decay of organic residues produce ammonium sulfate or other nitrogenous compounds that provide nutrients for plant growth, excessive concentrations of sulfur dioxide and nitrogen oxides from the burning of fossil fuels and from automobiles

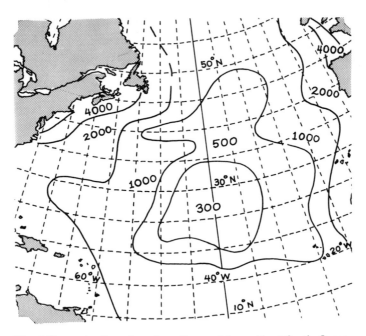

Fig. 5 Concentration of condensation nuclei over the Atlantic Ocean. (after Hogan)

produce sulfuric and nitric acids, chemicals highly corrosive and destructive to buildings, vegetation, and human lungs. This type of air pollution interferes with the ecosystem by preventing, delaying, or accentuating the normal phenomena that over a period of many centuries have produced the climatic pattern to which the regional organisms have adapted.

Increases in particles on a large scale over several decades might modify long-established weather patterns so that the global ecosystem could be quite different. While the restorative mechanisms of the atmosphere are massive, human beings, with their tremendous technological power, are capable of matching or even surpassing some of the capabilities of the natural world, especially in the production of airborne particles. They might thus also inadvertently alter the weather patterns that have evolved over a long period. If this were to occur in such a way that new cloud patterns developed on a continental scale, weather systems could be unintentionally modified to a degree not now predictable. The remarkable "year without a summer" of 1816 is thought to have been caused by massive volcanic eruptions and is an indication of what

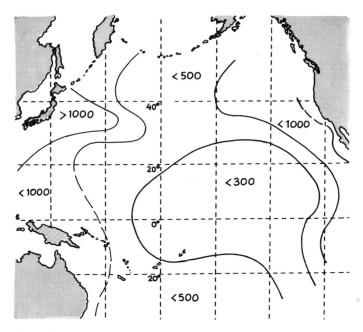

Fig. 6 Concentration of condensation nuclei over the Pacific Ocean. (after Hogan)

could happen — volcanic ash particles serve as excellent nuclei for ice crystal formation. This factor, plus the reduction in solar radiation caused by the volcanic dust cloud in the stratosphere and upper troposphere, is thought to have been responsible for the widespread change in the weather America and Europe experienced at that time.

Sampling Procedure

Airborne particles can be measured in a variety of ways. The larger particles are most easily sampled using a procedure in which several cubic meters of air are either passed through an air filter or directed at a sticky slide at a velocity sufficient to make the particles strike its surface. A velocity of 20 meters per second or higher will capture most particles of 5 microns or larger on a 2.5 × 7.5-cm glass slide with a sticky coating. The field procedure is described in Chapter 10.

Particles smaller than 1 micron may also be captured by an air filter. Because particles that are 0.1 micron or smaller weigh only

0.000,000,000,000,001 gram, a tremendous quantity of air must be filtered to obtain a sample large enough to weigh. A much easier method is available for measuring the *number* of small particles: The air sample is passed into a small chamber containing saturated water vapor. The sample air is then suddenly expanded into a vacuum chamber; as the air expands, it becomes supersaturated so that the airborne particles serve as nuclei for water condensation. The opacity of the fog is then measured with a light and a photocell, and the response is calibrated to indicate the number of particles contained in a cubic centimeter of the air sample. A simple version of the process is described in Chapter 10, and a description of the instrument is given in Chapter 7.

When measurements of airborne particles are made in various parts of the United States, the results appear as shown in Fig. 7. The "natural" regions typified by the mountains of northern Arizona and the forests of northern New York show less than 2000 particles per cubic centimeter. The areas affected by man's activities, such as in eastern Michigan and southern California, have particle concentrations 50–100 times greater.

Atmospheric Cleansing Mechanisms

Were it not for the cleansing mechanisms in the atmosphere, the amounts of gases and particles introduced by human activity (and sometimes natural phenomena) would soon become excessive and make most global air, especially in the mid latitudes, a serious health hazard. Fortunately there are several physical reactions that serve to cleanse the atmosphere.

When the concentration of airborne particles exceeds about 100,000/cc, rapid coagulation occurs between neighboring particles, especially if they are smaller in diameter than about 0.5 micron. These particles exhibit strong Brownian motion, i.e., they move continuously in random directions. This is caused by the bombardment of the tiny airborne particles with molecules of the atmospheric gases. (It is an interesting phenomenon and can be seen by illuminating a cloud of these particles, as from a match immediately after the flame is extinguished, using a low-power microscope with a magnification of 10–50X. The smoke should be viewed at right angles to the light beam and against a dark background.)

Particles larger than 0.4 micron never reach a concentration as high as 100,000/cc in the free atmosphere. It is the smaller ones that are highly concentrated and thus coagulate by collision with one another.

Coagulation is only one of the reactions that remove particles from the air. The second in importance occurs as clouds form, with most of the particles larger than about 0.1 micron acting as cloud condensation nuclei. As noted above, most of these particles start

Pl.6 The reflection of the sun from a myriad floating ice crystals. An undersun is seen in the direction of the sun at the same distance below the horizon as the sun is above the horizon.

acquiring water molecules before the air becomes supersaturated by about 1%. If some of them are of sea salt or other hygroscopic material, they may start accepting water at relative humidities as low as about 70%, some even lower.

Once moist air reaches its dew point, the cloud condensation nuclei grow rapidly until they reach a diameter of 10 to 20 microns. This depends, of course, on the number of water molecules and cloud nuclei available. The larger the number of these nuclei, the smaller the final water droplet diameter for a given atmospheric moisture supply.

As soon as a cloud droplet forms, it serves as a gathering place for the smaller fine particles in the air. Often, especially in polluted air, thousands of fine particles will contact and stick to a single cloud droplet. These fine particles have so little mass that once they contact the cloud droplet, they adhere to its surface or sink into its depths and dissolve. At times, especially when a stagnant air mass having high levels of pollution contains stratiform clouds, enough coalescence occurs that mist-sized droplets finally develop and fall from the cloud. These droplets have diameters ranging from 100 to 500 microns (0.004–0.020 in.). When collected on a glass slide and examined under a microscope after the water has evaporated, these droplets will be seen to have gathered so much residual matter that they are nearly opaque.

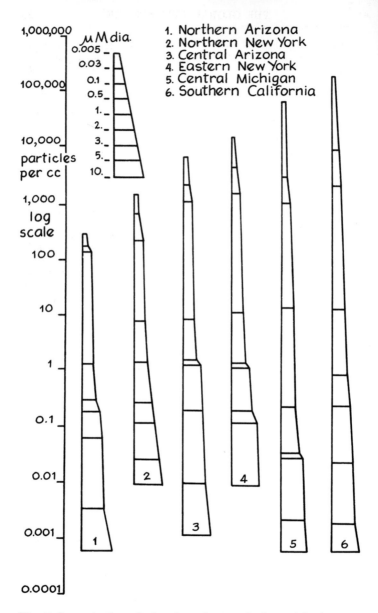

Fig. 7 Concentration of nine sizes of atmospheric particles in very clean air and with increasing levels of pollution.

When precipitation forms by coalescence, all of the particles that may have diffused to the cloud droplets are gathered into the liquid precipitation elements. As these fall, a great many of the microscopic and submicroscopic particles that were in the air are thus removed and again reach the surface of the earth and oceans where they originated.

Sometimes the cloud droplets become supercooled, and then evaporate when an ice nucleus starts growing in their proximity. The particles captured by such cloud droplets sometimes are brought together and remain as an aggregate that is much larger than the droplet first formed on a single cloud condensation nucleus. This aggregate may be intercepted by a falling snow particle. Fine particles not captured by contact with a supercooled droplet may be driven to a nearby ice crystal by invisible forces that are generated by the higher vapor pressure of the water droplet compared to that of a nearby ice crystal. The water vapor molecules streaming from the cloud droplet toward the ice surface may impact and push these fine particles toward the ice crystal, where they are captured on contact.

The final cleansing mechanism has already been mentioned — that of the capture of particles that may be directly in the fall path of raindrops or snowflakes (as described in the coalescence mechanism of precipitation). Only the larger particles are likely to be captured in this manner since the smaller ones move around the precipitation element in the streamline flow.

Thus it is that cloud formation and precipitation constitute the primary mechanisms for cleansing the atmosphere. It is important to realize that the atmosphere must be slightly "dirty," i.e., must have adequate numbers of cloud condensation nuclei and ice nuclei. Otherwise the functioning of the natural ecosystems of the planet would be badly upset. As the cleansing mechanisms are continually pushing the concentration of particles toward the so-called global background, an equilibrium is established between particle formation and removal based primarily on natural phenomena.

As the condensation nucleus level drops to the range of 500–3000/cc and the cloud condensation nuclei to 100–300/cc, the various cleansing mechanisms become less effective and the air is clear with almost unlimited visibility.

As these concentrations increase, and the ratio of condensation nuclei to cloud condensation nuclei rises from less than 10:1 to more than 30:1, the precipitation and coagulation mechanisms change. Submicroscopic particles grow so that they scatter light; visibility is reduced; and air pollution becomes a noticeable factor in the atmosphere.

In regions where human activities tend to overwhelm the natural cleanup mechanisms, the Clean Air Act and similar actions now beginning in other parts of the world are of great importance.

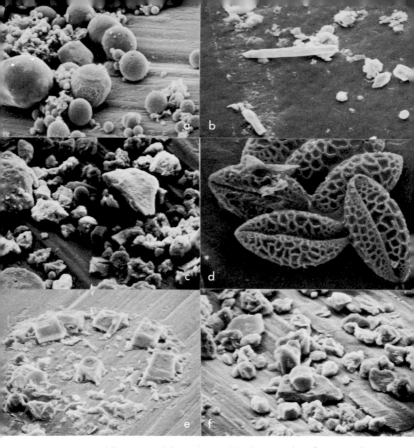

Pl.7 Large airborne particles as seen with the scanning electron microscope: a. fly ash from coal; b. asbestos from quarrying; c. clay soil; d. pollen grains; e. salt crystals; f. quartz sand. 10–50 microns.

That this program has had considerable success in the U.S. is readily evident, especially to the air traveler. The number of visible plumes from chimneys in the vicinity of major cities has been reduced to nearly zero. This is especially noteworthy for the air traveler who is interested in wind flow and related phenomena. Except for the direction in movement of cloud shadows across the earth, the landing pattern followed by the pilot, or the movement of dust clouds whipped up by the wind or car traffic on gravel roads, there is little indication from local pollution sources of wind directions at the surface of the earth.

Airborne particles are discussed further in Chapter 8.

Cloud Photographs from Satellites

With the development of modern rockets and the satellites and space platforms put into orbit, a new dimension has been added to the observation of the atmosphere, its clouds and the precipitation they produce.

Since satellite photographs depict clouds from altitudes of hundreds or thousands of miles, the viewer who is more familiar with seeing clouds from the earth, and thus at distances not exceeding about 30 km (50 mi.), must reorient past experiences to a new dimension of the atmosphere. The satellite photographs in this section are arranged so that those produced by the lower satellite are encountered first, and will be somewhat similar in appearance to those seen from the window of an airplane (as shown in Chapter 5). They are followed by photographs obtained at higher altitudes and then finally the remarkable views of the entire Northern and Southern Hemispheres made by a mosaic of a sequence of overlapping pictures. These sequences are now prepared on a daily basis for weather studies.

The spectacular cloud-structure signals taken from satellites and relayed to earth were anticipated in the mid fifties by photographs taken from jet aircraft flying in the lower stratosphere. Even before this use of high-altitude aircraft became common, flights of a number of V-2 rockets, fitted with cameras and launched in New Mexico in the early fifties, provided a preview of the fascinating cloud patterns that are now routinely measured by satellites located above the troposphere.

As satellites became a practical development in the late sixties, a series of them designed for weather monitoring were orbited: first the Tiros, next the Nimbus series, and then the Application Technology Satellite (A.T.S.). The A.T.S. is positioned at an altitude of 35,880 km (22,300 mi.) so that it remains stationary over a fixed point on the earth's surface. The Nimbus satellites, on the other hand, travel around the earth in about 110 minutes. They are in polar orbit, with the earth spinning underneath; so their sensors are able to complete global coverage in 24 hours. Thus the daily patterns of the planetary waves, cyclonic storms, hurricanes, and other large-scale weather patterns are under constant surveillance. This information is highly useful for short-range weather forecasting. The responsibility for monitoring these satellites has been assumed by N.O.A.A., the National Oceanographic and Atmospheric Administration.

The weather and earth reconnaissance satellites are equipped with highly sophisticated telemetry devices that relay signals to ground stations. These are then used to produce photographs. By obtaining simultaneous photographs of the earth at 5 or 6 specific wavelengths, remarkable variations in signals are obtained that can be interpreted as differences in radiative temperature of the various surfaces. This provides a wealth of geophysical information that, when properly interpreted, helps to differentiate between cirrus clouds, lower and warmer clouds, snow, lake, and sea ice, and such special features as the boundaries of the warm waters of the Gulf Stream which often produce clouds when the colder air from the Polar regions flows over them.

The A.T.S. satellites also provide the time-lapse cloud photographs now used on the television weather programs. The original maps produced by satellite are far better than those shown on the TV screen, so improvement in the picture quality of these presentations can be expected.

In addition to the satellites just described, from which most of the photographs reproduced in this section of the guide were taken, another satellite series, the Earth Resources Technology Satellite (E.R.T.S.), launched July 23, 1972, provides a view from 435 km (263.7 mi.) of every portion of the earth. Photographs from this satellite are readily available to the public at nominal cost from EROS Data Center, Data Management Center, Sioux Falls, South Dakota 57198.

Readers who are interested in their local cloud, storm, and terrain features should visit one of the regional "browsing" offices of the U.S. Geological Survey. There they can review the variety of photos that are available. Individual photographs can be obtained that cover an area of approximately 26,569 km^2 (10,096 mi.^{2}). They are thus highly useful for providing information about the types of clouds and storm systems that occur over a specific region.

Satellite photographs of clouds, and the weather patterns of which they are a part, show that the clouds encountered daily are all part of a continuous global system. They are not random occurrences but rather are part of an energy interplay between solar radiation, nighttime cooling, the seasonal changes, and the pressure patterns that develop from these interactions. The cloud patterns that are seen develop from large- and small-scale rising air motions that bring about cooling and cause tongues of moisture to undergo condensation.

The incessant rising and sinking of the atmosphere produces the intricate patterns of cloud and noncloud whose interpretation challenges the best scientific minds.

Pl.8 Remarkable pattern of convective clouds forming curved lines in cold air as it moves from land across the open ocean. Center of the vortex is over Gulf of Maine. 12/17/72 NOAA 2

Pl.9 Extensive sheet of altostratus near vortex of large cyclonic circulation. Cirrus clouds, mostly of cirrostratus type, cover a large portion of the storm; in some areas it is quite diffuse. 9/4/73

Pl.10 View of Hurricane Ava over the eastern Pacific. Eye of storm shows it to be the center of a huge vortex with spirals of cumulus and altocumulus moving toward center. 6/7/73

Pl.11 Large storm center over Labrador showing highly organized vortex with extensive areas of altostratus and altocumulus (as billows) with cirrus above and closest to the center of the storm. NOAA 2

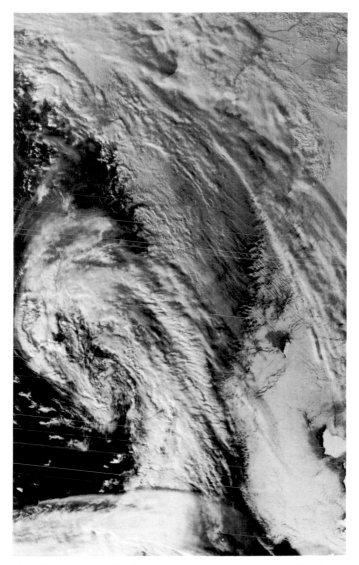

Pl.12 Clouds over the Bering Sea. Ragged pattern on right side of view is the ice pack. Cloud of cirrus and altocumulus has formed over open water and shows strong cyclonic vorticity. 3/4/73 NOAA 2

Pl.13 Snow pattern resulting from intense blizzard covering extensive area south of Hudson Bay. Cirrus in area west of Great Lakes is part of new storm that is forming. 4/11/73 NOAA 2

Pl.14 Lake-effect snowstorm underway on lower side of Lake Erie. Snow falls under the narrow cloud bands extending more than 80 km (50 mi.). 1.2 m (4 ft.) of snow may fall in local regions.

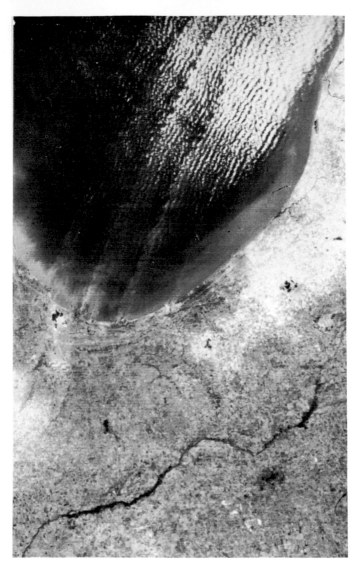

Pl.15 Plumes from industrial plants along south shore of Lake Michigan seem to influence location and wavelength of bands of altocumulus lines of clouds (cloud streets) parallel to wind. 11/24/72 ERTS

Pl.16 Clouds forming over ocean in westerly flow from cold land that, in infrared filter, is white. Blackest portion of ocean shows the warm waters of the Gulf Stream. 11/18/72 ERTS

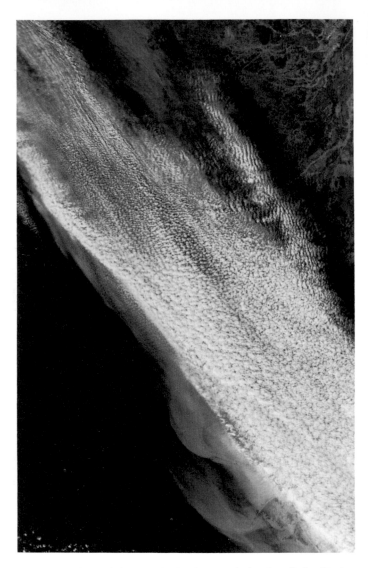

Pl.17 Long sheet of altocumulus having a variation in cell size. Notice how cloud elements are in orderly array in several directions. Ripples indicate location of shear.

Pl.18 Convective clouds over the Atlantic adjacent to the east coast of Delaware. Cold, cloudless air over land leads to long lines of clouds over the moist ocean air above warm water. 12/17/72

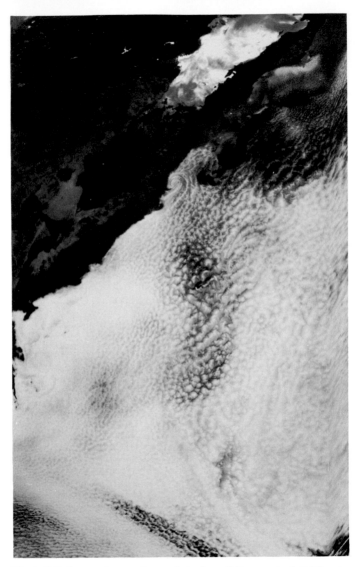

Pl.19 Cellular stratocumulus and stratus off the coast of California. Note the eddies in the clouds adjacent to the land in upper center: clouds were not formed over dry land. 4/28/73 NOAA 2

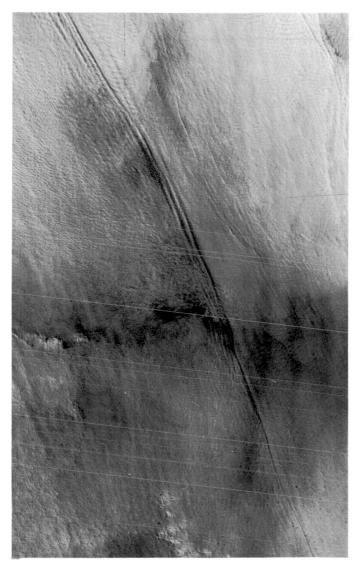

Pl.20 An extensive sheet of altostratus with a long line of gravity waves that probably mark a shear zone. Other less noticeable waves are present along with very small ripples at right angles.

Pl.21 The long curved arc of a jet stream passing over a lower deck of stratus covering an extensive portion of the Mississippi Valley. Ribs of cirrus extend northward from edge. 1/8/73 NOAA 2

Pl.22 A variety of clouds over the western Atlantic Ocean near Greenland. Cirrus is dominant over the upper portion with some large cumulus between it and rippled altocumulus. 1/1/73 NOAA 2

Pl.23 A complex mixture of cirrus clouds associated with a jet stream south of Hudson Bay. Much of northern area has clouds thin enough to permit land forms to be seen. 12/10/72

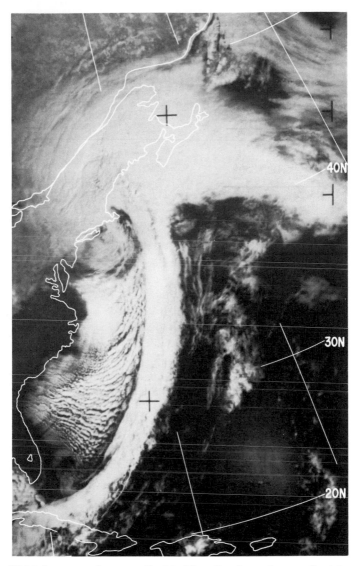

Pl.24 A wave cyclone over the Maritime Provinces. A warm front to the east with a cold front trailing to the south. Notice the cloudfree area southeast with cloud over the water. 3/4/71 ESSA 8

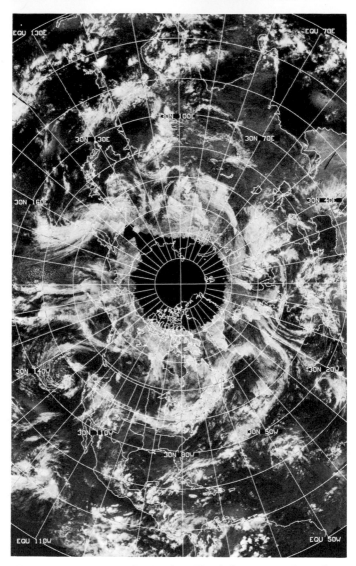

Pl.25 A satellite view of the Northern Hemisphere as seen from above the North Pole. Streamers of very cold air flow southward to form cloudy spirals in the middle latitudes. 10/12/72

44

2

Clouds, Air, Condensation, and Atmospheric Motion

A wealth of information about present and forthcoming weather is displayed in the sky, but the language of clouds must be learned before it becomes meaningful or useful.

In the past the mariner, the shepherd, the farmer, and the miller learned how to read the sky because their livelihood, and sometimes their safety, depended upon their ability to forecast sudden changes in the weather. This ability is less cultivated today, since use of electronic communications has greatly speeded up the transmission of weather information to everyone who needs it.

The sky also has aesthetic value, and the beauty of nature would be greatly diminished if there were no clouds to provide a varying panorama of form and color.

One of the main problems of sky-reading is changeability. The clouds are never the same in form, structure, and color, and their messages often do not remain in the sky long enough to be pondered.

Another problem is that of complexity; in most instances the message is not simple, and there may be several messages superimposed on one another. Yet, the 20th-century scientific orientation suggests that there is a logical answer, and that there are causal relationships underlying both the complexity and the simplicity of the sky.

This confidence in a rational explanation has not always existed. Prior to 1800, observers of the sky spoke of clouds only as a vague category of "essences" floating in the air; clouds had no names, nor were they understood. Many other aspects of the atmosphere were only beginning to be explained scientifically, and atmospheric science was still in its infancy.

One young Englishman thought independently about clouds and introduced a whole new perspective. An astute observer of the sky for many years, Luke Howard (1772–1864) felt there must be a cause-effect relationship operating in the atmosphere to produce the many different forms of clouds.

In his late twenties Howard belonged to a small club of scientifically oriented Londoners, the Askesian Society. In the winter season of 1802–3, he presented a paper that became a classic, "On the

Modification of Clouds." (Today, "modification" would be better expressed as "classification.") As part of this new classification system for clouds, Howard developed a set of Latin names for the main and secondary cloud types.

Howard's paper was so perceptive that now, nearly 2 centuries later, the classification system and cloud names are still the same as his in some respects and closely resemble them in others.

Luke Howard's practiced and keen eye discerned the presence of 2 primary categories of cloud, *heap clouds* and *layer clouds,* and these are still the basis of most classification schemes. They also will be the 2 primary classes discussed here, identified as Group I and Group II. There are 3 additional groups, as shown in Fig. 8. Four groups are discussed in detail here.

Group I. Heap Clouds — The Cumulus Family

These are flattish-based clouds of domed cauliflower shape, with sharply defined edges and the tops of towers rising to different

Fig. 8 Cloud Classification.

GROUP (av.)	Tier or Layer		
	Low (0-3 km)	Middle (3-7 km)	High (> 7 km)
I (HEAP)	cumulus fair-weather	swelling cumulus	cumulo-nimbus (non-precip.)
II (LAYER)	stratus	alto-stratus	cirro-stratus
III (LAYERS ę HEAPS)	strato-cumulus	alto-cumulus	cirro-cumulus
IV (PRECIPITATING)	nimbostratus cirrus cumulonimbus		
V (UNUSUAL)	separate listing (note: 1 km = 3280 ft.)		

heights (see Fig. 9). The vertical and horizontal dimensions of the heap clouds are approximately the same. On some occasions, the vertical dimension is only a few hundred meters while on other occasions the cloud tops rise to heights of 20 km (65,600 ft.). The horizontal diameter ranges from that of a city block to a few kilometers. Individual clouds are well separated from each other. Cloud color is usually white, though it can change to dark, threatening gray depending on form of development, illumination, and age of cloud. The cumulus, or heap cloud, is a convective cloud whose essential character is that of many rising bubbles of relatively warm air. If moisture is evenly distributed horizontally through the local air mass, the ascending air generally reaches its saturation point at the same level. The onset of condensation produces the flattish base (which is reported by glider pilots to be slightly higher at the *center* than around the edge). Descending air in the surrounding region experiences heating and drying and causes the cloudless space.

Plate 26 shows a sky of fair-weather cumulus (Group I_L) in vary-

Fig. 9 Group I. Heap Clouds—The Cumulus Family.

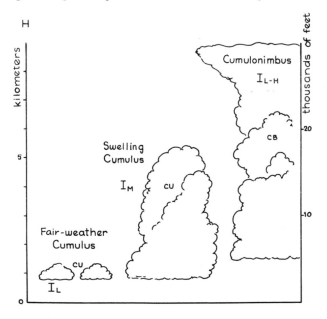

Pl.26 1 shows the flattish bases of the clouds, corresponding to the lifting condensation level. The tops of the clouds, **2,** resemble small cauliflowers, the rounded elements indicating the presence of a great many small cells of convective activity; the cells may be at any given stage of development (each has a lifetime of perhaps 10 minutes). **3** shows a cloud edge in the process of evaporating and taking on a frayed appearance. **4** is the remnant of a cloud mass that has gone through its life cycle, expended its energy, and is now sinking and disintegrating into fractocumulus, or broken-up cumulus. **5** shows clear areas in which the predominant motion is downward, compensating for the upward motion indicated by the presence of cloud.

48

ing states of development. This is a settled-weather cloud form. It usually appears in late morning or early afternoon, produced by upward-rising convection currents stimulated by contact of the surface air with the sun-warmed earth. Warmed bubbles or "blobs" of air break away from the earth and rise, seeking to find a level of temperature equilibrium. Ascent brings about cooling, an increasing degree of saturation, and eventually condensation. Heat energy carried in the water vapor ("latent heat") is released when condensation takes place. The convection continues, but the generally settled state of the atmosphere limits vertical development to only a few hundred to a thousand meters (1 km, or about 3000 ft.).

Pl.27 The swelling cumulus (Group I_{LM}) shown here represents a more energetic and therefore more unstable state of the atmosphere, which allows the convection cells to rise to higher altitudes. An individual cloud mass is seen on close inspection to consist of many cells that in turn are composed of still smaller cells, all at different stages of development. The general air movement in the cloud mass is upward, with compensatory downward movement around the perimeter, and local regions of sinking air within the general cloud mass. As with fairweather cumulus, bases are generally flat, **1.** A well-developed upward-moving chimney-shaped cell is seen at **2.** The cauliflower shape now becomes domed towers, **3.** Wisps of fractocumulus cloud are sometimes found around the cloud mass, as at **4,** representing small residues of condensed vapor. **5** shows the beginning of a new cloud mass. At **6** the vertical motion in the interior of a cell is now swinging out and downward, causing cloud elements to begin evaporating. This particular cloud is quite high, with bases a few thousand meters above Mt. Agassiz (in n. Arizona) which is itself 3780 m (12,600 ft.) high. Therefore, the cloud droplets are probably supercooled.

49

Pl.28 The cumulonimbus (Group I_{LH}) shown here represents a still more energetic state of the atmosphere. The vigorously growing cloud mass consists of hundreds of thousands of individual convective cells. Large amounts of heat energy are released as the upward-moving water vapor condenses into liquid form. This has the effect of making the rising air more buoyant, as happens with a balloon when hot air is injected into it. Starting from the dark bases shown at **1**, we see the cloud tops at **2** still pushing upward.

The numbers identify the upward extremity of an organized convective cell that consists of many smaller cells. The greatest activity is shown at **3** and **4**. **5** shows a region in which there may be relative sinking motion. At **6**, in the rear of the cloud, there is a suggestion of glaciation (conversion to ice), one of the marks of a transition from pure cumulus to cumulonimbus. This cloud can therefore be considered a very young cumulonimbus.

Group II. Layer Clouds — The Stratus Family

Two features distinguish members of this family from those in Group I (p. 46). The first is that they appear in layers with the vertical dimension small compared to the horizontal, like a sheet of paper or a mattress. Whereas the thickness of a typical stratus cloud may be 0.5–1 km (about 1600–3200 ft.), the horizontal dimension may range from 10 km (6 mi.) to 1000 km (600 mi.). The area covered by the cloud can be 1,000,000 sq. km (360,000 sq. mi.), which is larger than the area of the state of Texas. The second difference is the absence of convection cells.

The cause of the expansional cooling that in turn produces the sheet of cloud is the gentle ascent over another heavier mass of air, or over a gently sloping land form, or horizontal squeezing (convergence) and compensatory uplift in the air-mass flow.

The cooling that causes clouds can be brought about in another way. When the ground loses heat during the night, the air in contact with it is cooled. When there is enough moisture in the air, and there is little or no wind, the air temperature may be lowered to the point at which dew forms (the dew point temperature) and condensation will occur. If there is just enough air motion to produce gentle stirring, so that additional air can be cooled by mixing or direct contact with the cold ground, the condensed moisture — fog — can attain a thickness of 100 m (300 ft.) or more. When the sun rises and the earth warms, the air temperature will again rise above the dew point and cause the fog to dissipate. Sometimes fog occurs as a rather local phenomenon; it can also be widespread, as when the slow upglide movement of air from the Great Plains toward the Rockies brings air to its dew point and several states are blanketed in persistent fog.

Fig. 10 Group II. Layer Clouds—The Stratus Family.

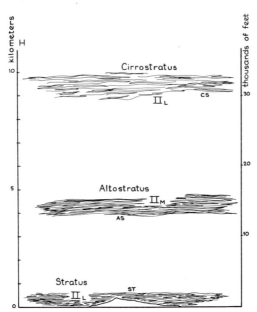

Pl.29 Group II$_L$ clouds in the stratus family that lie on the ground are called fog. The photograph shows ground fog, **1,** from within the cloud. The cloud has no form. It is midmorning and the cloud is thin enough for the position of the sun, **2,** to be identified.

52

Pl. 31 Stratus cloud in the middle layers (Group II$_M$) is called altostratus. The middle elevations are thought of as roughly between 3 and 6 km (10,000–20,000 ft.). Altostratus is primarily a cloud produced by upglide motion over a warm front. It is predominantly a water-drop cloud, though the water drops may be supercooled. General appearance is darkish gray. Sometimes it is thick enough to obscure the sun; at other times, when ice crystals are present, the sun looks as it does when seen through a piece of ground glass. In a storm, altostratus tends to thicken into nimbostratus from which precipitation falls. Plate 31 shows a rather typical altostratus cloud layer, or deck, exhibiting the ground-glass effect, **1,** and a dull gray appearance, **2.**

53

Pl. 30 (*left*). Some low stratus, known as advection fog, is formed when an air mass moves horizontally over a colder surface and is cooled until it reaches the dew point. The fog characteristic of the summer season off the California coast and in the inland basins (e.g., San Francisco Bay) is of this type. It has a low ceiling, and its top is limited by a temperature inversion. The photograph shows this fog pouring through the Golden Gate Bridge from the Pacific Ocean, **1,** into San Francisco Bay, **2.** The time is late afternoon. Again, the fog is an essentially featureless cloud, though the top, **3,** is more sharply marked than the base, **4.**

Pl.32 A uniform pattern of cirrostratus, **1,** is spread across all parts of the sky. The cloud appears thicker in the lower portions of the photograph, but this is an illusion due to perspective.

54

Group III. Heaps and Layers

The presence of stable layers in the atmosphere converts vertical into horizontal motion. This happens so frequently that it deserves this separate group identification.

At the lowest level, layerlike stratiform clouds frequently occur. These show evidence of convective cells, which produce thick and thin regions in the cloud, marking the rising and sinking motions respectively. This cloud is called *stratocumulus*.

In the middle layers, *altocumulus* has the same cellular composition. Sometimes the cells are aligned in rows. Generally speaking, the cloudless regions are much smaller than the clouded space occupied by the rising air.

At the uppermost levels, higher than the freezing level, the same pattern persists in a cloud composed of supercooled water and some ice crystals. The individual convection cells look smaller but this is mainly due to their being at a greater distance from the observer. This cloud is the *cirrocumulus*.

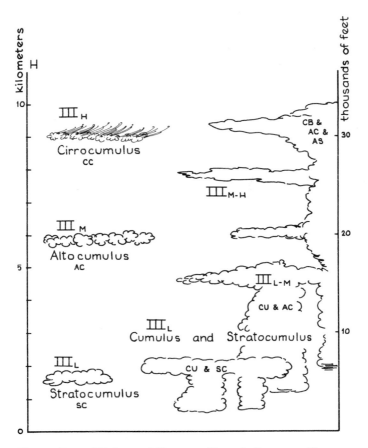

Fig. 11 Group III. Layered Heaps, or Heaps in Layers, or Heaps and Layers.

Pl.33 Illustrated here is stratocumulus (not formed by the flattening of cumulus tops). Bases, **1,** are at uniform heights. Any apparent foreground lifting of the base is due to perspective. Open spaces, **2,** indicate the presence of the descending portion of the convection cell. **3** points out the cloud tops. Maximum cloud thickness is estimated as about 0.3 km (1100 ft.). Bases might be at 1 km (3200 ft.).

Pl.35 (*right*). Shown here is the beautiful cirrocumulus (Group III_H), a cloud composed mainly of supercooled water droplets with some ice crystals. This pearly white cloud with closely spaced, puffy elements usually appears with cirrus, which in this case is shown by the ice crystal streamers falling from the cloud. The cloud usually has structure and pattern. Here the globular elements are lined up as lower-left to upper-right rows at **1** but change orientation at **2.** The most dense center where cirrocumulus and cirrus are mixed is at **3. 4** shows the cirrus edges of the cloud; **5** is a region where the row alignment has been lost or has not yet been established.

Pl.34 (*above*). Shown here is altocumulus (Group III$_M$). **1** identifies puffy regions in the cloud in which air is rising. Sinking, **2**, is marked either by open space or thinness in the cloud. Note also that there are two row alignments roughly at right angles to each other, one from the left to right, another larger one from topright to bottomleft. Wave structures in altocumulus are common features, and more examples will be shown under Group V.

57

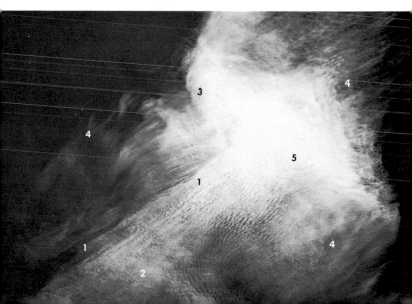

Pl.36 There are many further combinations of heaps and layers, resulting from the attempt by the individual heap cloud to grow vertically and yet remain separate from others, and the restraint put upon this attempt by the presence of the stable layer that says, in effect, "You can't go through but you can spread out." The factors that determine the appearance of a particular sky are: (a) height of the lifting condensation level, (b) strength of convective activity, (c) height and strength of the stable layers.

These possibilities are illustrated here. There is strong inversion, **1,** at about 2 km (6500 ft.) with the lifting condensation level, **2,** at 1 km (3300 ft.). In this unusual instance the limiting of upward cloud motion is shown in convenient cross section. Small cumulus towers, **3,** grow as at **4** and on reaching the inversion level, **1,** can only spread out. The sky then becomes a grouping of small cumulus and stratocumulus.

Group IV. Clouds that Generate Precipitation

Nimbus is the traditional term for clouds of this type. The most common precipitating cloud found in low and middle layers is the *nimbostratus*. In general, it has a darkish gray appearance, ragged edges, and rain or snow falling continuously from its base.

In the winter season, at middle and high latitudes, the freezing level often comes down to the surface, with cirrus-type clouds found at all elevations. It is not uncommon under such circumstances for a cirrostratus layer to produce a gentle "rain" of columnar ice crystals that fall to the ground before they have a chance to evaporate.

One important factor affecting all clouds is the amount of water vapor the atmosphere contains. The standard way of specifying water content of clouds is in terms of water-vapor density measured in grams per cubic meter (g/m^3). A gram of liquid water occupies very nearly the volume of 1 cubic centimeter, which is conveniently remembered as the size of a small sugar cube; a cubic meter is about the volume of an ordinary desk. The average values of water-vapor density in clouds range from about $0.1\ g/m^3$ to $3\ g/m^3$.

Fig. 12 Group IV. Clouds Generating Precipitation.

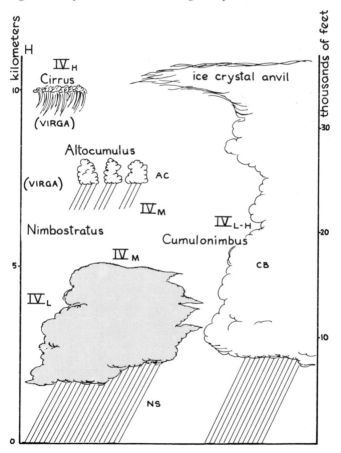

Pl.37 Illustrated here is the nimbostratus cloud (Group IV$_{LM}$). The dark gray sky, **1**, produces regions of precipitation, **2**, often accompanied by ragged storm clouds, **3**, known as scud.

When a cumulus or cumulonimbus cloud generates a shower or squall from its base, it is identified as a nimbus (Group IV) cloud.

Clouds are dynamic entities, and they may change from one form to another. This is particularly the case in the cumulus family; at different times of observation the same cloud might be classified as I$_L$, I$_{LM}$, I$_{LH}$, and finally as IV$_{LH}$.

Pl.39 (*right*). The cumulonimbus cloud illustrated here is massive. It is the only cloud in the sky, and its base, **1**, is so far away that it lies below the horizon. Its top is estimated as 18 km (60,000 ft.) and its diameter between 15 and 20 km (47,500–68,600 ft.). In the case of this cloud, a time-lapse film run at normal speed showed it in ponderous clockwise rotation about a vertical axis, a most impressive spectacle!

Though the lifetime of an individual convection cell is recorded in minutes, this giant cloud's life was perhaps 12 hours, and hundreds of thousands of individual cells participated in its growth.

The cloud shows the effect of a definite lid imposed by an extremely stable layer (inversion), which is probably the base of the stratosphere. Prevented from further vertical motion by the inversion, the updraft is diverted laterally and produces the shelf structure seen at **2**, where stringiness at the cloud edge shows glaciation. The tremendous activity in the cloud is shown by the roll structure at **3**. Examples of the many cells are shown at **4**. Virga falls from the cloud at **5**.

At times the updraft velocity within a giant cumulonimbus cloud is so strong that its upper portion is injected into the stratosphere.

Pl.38 (*above*). A classic precipitating cumulonimbus is illustrated here. **1** shows the widespread flattish base. The top of the cloud is obscured by the clouds at **2** but must penetrate well into the high troposphere. **3** shows examples of the many active convective towers associated with this large cloud. The heavy squally precipitation, **4,** falling from the base indicates the cloud is at full maturity. Other nearby clouds, **5** and **6,** are at a somewhat earlier stage of development, while the ragged edges of the cloud at **7** indicate it is sinking and disintegrating.

61

Pl.40 Ice crystal clouds that present themselves in varied shapes and forms in the colder-than-freezing levels are called cirrus. Except in the winter season, these clouds are found at high elevations. As the growing ice crystals become larger they get heavier and fall earthward. Though the streams of falling ice crystals generally evaporate while still in the high atmosphere, such clouds are classified as belonging to Group IV because they are generating precipitation. The strange, distorted forms taken by the wisps of falling crystals that constitute the cirrus cloud result from the differing wind speeds and directions in the various lower layers.

Shown here are the principal generating centers, **1,** of the ice crystals and the trails, **2,** of falling crystals. The regions of thick and thin cloud are related to the relative activity of the precipitation process. We infer that the high-level winds are blowing from left to right, with upper winds stronger than lower, thus giving the impression that the lower trails of precipitation are left somewhat behind.

Are natural atmospheric processes well enough understood to explain the phenomena that are observed? The answer is a qualified yes. Atmospheric scientists (including cloud physicists, aerosol chemists, and meteorologists) understand a great deal about how clouds and precipitation are produced. However, they admit that their knowledge is incomplete.

If a phenomenon is to be explained scientifically, it must first be described. Understanding is always limited by the completeness and accuracy of description that is possible, and this is one of the sticky points of atmospheric science. The atmosphere is a tremendously complex system, and difficult to describe in complete and accurate detail, although great progress is being made in that direction.

On the following page is a visual summary of the precipitation process, the so-called precipitation "staircase." This is called a staircase because it separates the series of steps that must follow, one after another, in the formation of a cloud and its precipitation.

Composition and Behavior of Air

The substance that is called air is a mixture made up of different gases, some plentiful and others very rare, continually blended by atmospheric motions; by volume, it is about 78% nitrogen, 21% oxygen, and nearly 1% argon, with fairly constant trace percentages of helium, ozone, neon, and krypton (other gases, including carbon dioxide, appear as variable trace gases). In the lower layers of the atmosphere — the troposphere — these gases are always found in the same proportions.

Most important, however, of the variable-percentage gases is water vapor; in equatorial maritime climates water vapor may be present in quantities ranging up to 7% of any given volume of the atmosphere. That is, 1 cubic meter might contain 1 tablespoon of water molecules. In arid desert regions, in the frigid Arctic, or at high elevations at the top of the troposphere, water vapor is still present, but in negligible amounts. The gases that make up our ocean of clean, pure air are a mixture that has evolved over a time span of millions of years.

The air found at the bottom of this atmospheric ocean is anything but clean. It is polluted by a variety of foreign substances that come from both natural and manmade sources. One of the primary natural mechanisms for introducing pollutants into the air is the frictional rubbing of the wind over the earth's surface. When this occurs over the land, surfaces of rocks are slowly abraded, and further breakdown in particle size comes about by the polishing effect of small particles rolling over their neighbors in somewhat the same way that agates are polished in a rock tumbler. The dust from this polishing effect is fine enough to be carried aloft by the wind, and it becomes more or less evenly distributed throughout the atmosphere. Dust storms, such as plagued the central states in the 1930s, are a dramatic example of this mechanism at work. A feature of considerable meteorological significance is that some of the particles are water absorbing, others are wettable but not soluble, and still others are water-resistant (hydrophobic).

Frictional rubbing of wind over a water surface is what causes waves. As the wind grows stronger, the height of the wave increases. When the wind speed exceeds about 40 km/hr. (25 mi./hr.), filaments of liquid are torn from the tops of the whitecaps. They disintegrate into spray droplets, most of which are so heavy that they fall back into the water; smaller ones, however, fall slowly and evaporate, leaving salt particles whose mass is a million millionths of a kilogram. These are the so-called giant salt nuclei of the atmosphere. A mechanism of great meteorological importance is the bursting, where sea and air meet, of bubbles of air that have been released from solution or have become trapped under the water surface. Jet drops and bubble film droplets result-

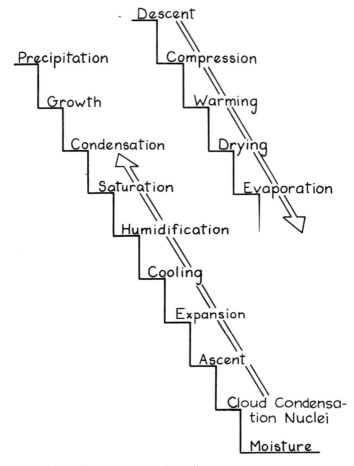

Fig. 13 The "Precipitation Staircase"

ing from the bubble-burst evaporate very rapidly in the free air because they are so small. They leave tiny residues of salt that become centers for future cloud droplets. A fourth important process that produces submicroscopic particles (much smaller than the fragmentation and evaporation residues just described) results from the gas-to-particle condensation described in Chapter 7.

An important natural mechanism that operates rather infrequently, but sometimes with a massive effect, is volcanic eruption, which injects immense quantities of dust into the high troposphere and the stratosphere. Other, secondary mechanisms are: (a) introduction of meteoric dust into the upper atmosphere as the earth moves in its orbit through dusty regions in space (like comet tails), and (b) the reaction of various trace gases and water vapor under the influence of solar radiation. Combustion by-products also pollute the atmosphere, as noted earlier.

Today scientists have attained a reasonable understanding of the various kinds of air pollutants, particularly in the larger size ranges. In addition to knowing what the pollutants are and how they are produced, scientists know their size distribution and their range of concentration. Perhaps most important of all, they are coming to understand the role pollutants play with respect to the water in the atmosphere.

Air Movement

Ascent The earth is a spinning globe. The earth's atmosphere is gravitationally attached to the globe and it is also spinning. Since it is not rigidly attached, it sometimes spins faster than the globe and sometimes lags behind. Air also moves up and down as it tries to achieve gravitational equilibrium, in very much the way that water, initially at different heights in 2 connected containers, moves to achieve the same level once the restraining barrier has been taken away.

The unequal absorption, by latitude, of solar energy over the surface of the earth is continuously upsetting the state of gravitational equilibrium, and the earth's atmosphere continuously moves to restore it.

Air ascends for a number of reasons. Fig. 14 summarizes the causes of rising air. The first and most obvious is that there are many physical barriers that the air cannot go through, and therefore must go over. There are hills, mountain ranges, and hummocks of all sizes and shapes.

Second, there are "hills" in the air itself, caused by differences in density. Colder air is heavier and more dense than warmer air. Consequently gravity causes it to hug the surface more persistently. When masses of air move from different regions on the surface of the globe and interact, boundary surfaces called "fronts"

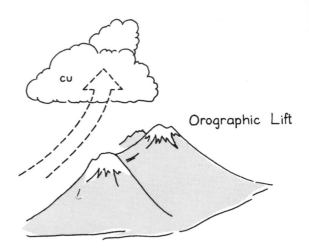

CU

Orographic Lift

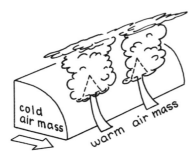

Cold Front

cold air mass

warm air mass

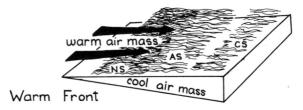

warm air mass

CS

AS

NS

cool air mass

Warm Front

Fig. 14 Cloud and front formation—causes of rising air.

are formed and move with resulting air motion over them, as illustrated in Fig. 14.

Sometimes horizontal flow patterns cause squeezing in the airstream somewhat like that which takes place when a freeway ends and the traffic in 5 lanes must fit into 2. Compensating for this horizontal squeezing in the airstreams is a flow upward; this condition is horizontal convergence. Its opposite, horizontal divergence, produces sinking motion in the airstream.

Finally, there is convection, explained by Archimedes' principle of buoyancy. A locally warmed, and thus less dense, volume of a fluid will experience a positive (upward-directed) buoyant force, moving upward and cooling until its density becomes the same as the density of the rest of the fluid at the same elevation. Because the fluid (air, in this case) rises, a compensatory sinking flow will appear around the rising column, as shown in the illustration.

The atmosphere is literally permeated with convection cells of all dimensions, ranging from those that are microscopically small to those that are global in scope. Convection cells are of great importance since convection is the process that transports energy from one place to another. In nature, excess heat energy must be moved from low-latitude to high-latitude regions. The earth's atmosphere and oceans are the means by which this transport is accomplished. Both fluids have an important function in this respect (see Fig. 15).

The 18th century Englishman George Hadley was one of the first to appreciate the global dimension of convection. The cell shown in the diagram is a Hadley Cell, and it represents the most basic circulation pattern, always present though sometimes obscured by the complexities of the real global circulation.

Expansion Atmospheric pressure decreases with elevation, but not in a simple linear way. Each time the elevation is increased by 8.4 km (27,560 ft.) the atmospheric pressure at the upper level decreases to nearly one third of its value at the lower level. Examples of precise values are shown in Table 2.

When air rises as a consequence of barrier, convection, or convergence factors, it finds itself under less pressure, and adjusts by expanding.

Cooling The atmosphere is a gas that is held to the surface of the earth by gravity. Table 2 shows that the pressure exerted by the atmosphere is greatest at the surface of the earth and decreases with elevation. As stated in Chapter 1, pressure decreases much faster in the lower levels of the atmosphere, so that about 50% of the atmosphere is contained in a paper-thin spherical shell that is only about 5-6 km (3.1-3.7 mi.) thick.

Air contains water vapor; how much it contains varies from day to day at any given place at any given time, and from place to place. As mentioned earlier, water vapor is the most variable component of air.

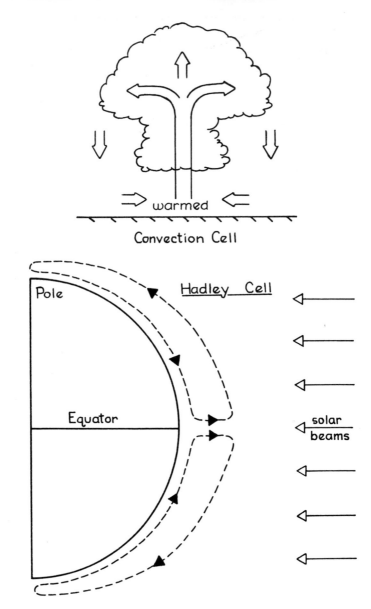

Fig. 15 Two examples of convection cells.

There is usually more water vapor in regions close to the surface of the earth, and particularly the ocean, since water vapor enters the air by evaporation from such sources as bodies of water and masses of vegetation. There is also more water vapor potentially in warm regions than in cold regions, since warm air can hold more water vapor than cold air; however, warm regions may also be very dry if cut off from sources of water — the Sahara is an example.

The law that covers the behavior of gases is called the General Gas Law, defined in elementary chemistry and physics texts. This law shows how a gas in one state, described by its pressure, density, and temperature, must adjust as it moves into a different environment and thus into a new state.

The atmosphere is restless, always in motion either horizontally or vertically or both. As air rises, pressure on it decreases and in response it expands. The act of expansion to encompass its new and larger dimensions requires an expenditure of energy; since temperature is a measure of internal energy, this use of energy makes its temperature drop, and the drop in temperature is 10°C for each kilometer of ascent. A mass of rising air — whether large and rising gently or small and rising abruptly — cools at this rate, known as the Dry Adiabatic Lapse Rate. Conversely, sinking air is compressed and warms at the same rate. This activity is of great meteorological significance: it determines the altitude at which a cloud will form or evaporate.

The average change in temperature relative to altitude that characterizes the global atmosphere by season and by latitude is shown on the next pages. Between sea level and 15 km (49,200 ft.) of altitude, the temperature is seen to drop by 28–56 Celsius degrees (50–100 Fahrenheit degrees), depending on the latitude. Then, between 15 km and 50 km (49,000–164,000 ft.), the tempera-

TABLE 2

Pressure (P)	Height (H)	
mb*	km	ft.
1013.2	0	0
372.3	8.4	27,560
137.1	16.8	55,120
50.4	25.2	82,680
18.5	32.6	106,960

* millibar (assuming temperature remains constant with height — a simplifying assumption)

ture either is constant or increases. Between 50 km and 80 km (164,000–262,000 ft.) it again decreases, following which from 80 km to 110 km (262,000–361,000 ft.) it increases rapidly. This behavior may seem unlikely, but that is not the case. Considering the distribution of water vapor in the atmosphere, the density of the atmosphere, and the way radiant energy enters and leaves the atmosphere, the curve shows just what would be expected.

The atmosphere is stratified in layers of varying degrees of Stable and Unstable Equilibrium; a mechanical analogy would be a cone resting on its base — it is in a state of stable equilibrium

Fig. 16 Representative Lapse Rates.

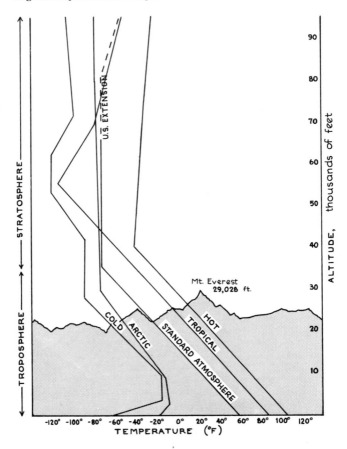

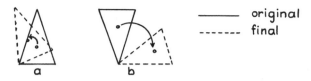

Fig. 17 A cone in a position of stable equilibrium (a) and unstable equilibrium (b). Center of mass raised in a, lowered in b.

because, if tipped, it will come back to the original position, as shown in Fig. 17. However, if it initially is precariously balanced on its tip, any slight nudge will move it to a new, lower position, as in **b.** Speaking in terms of the resulting movement of the cone's center of mass, the tipping in **a** lifts the center of mass and it returns to the same level. In **b** the center of mass falls to a lower, more preferred position.

The vertical motion of air is somewhat like this, as shown by the following examples (see Fig. 18). A hypothetical mass of air is lifted, expands, and cools at the adiabatic rate of 10°C/km. Level for level, its temperature is always less than that of its surroundings. In Fig. 18A, **a′** is colder than **a;b′** is colder than **b,** and the density is greater than that of its surroundings. It experiences negative, downward-directed buoyancy, and sinks. In cases like this the surrounding environment always acts to suppress vertical motion. Such a temperature stratification produces a state of equilibrium, and the lapse rate in the example shown is called a temperature inversion, since the temperature increases with elevation.

Conversely, if the atmosphere were differently stratified, as shown in Fig. 18B, the rising parcel would always find itself warmer (since **a′** is warmer than **a,** and **b′** is warmer than **b**) and less dense than its environment at the same level; this would give it positive buoyancy and would generate convective vertical currents. This condition is one of unstable equilibrium and is characterized by an absolutely unstable lapse rate. **a** and **b** represent extremes found in the atmosphere; there are many intermediate possibilities that occur more frequently and are found in different places and circumstances.

Humidification

Water is the most important substance in the process of life on earth. A very important insight, gained around 1800 by the English chemist John Dalton, was that each individual constituent in air (nitrogen, oxygen, and so on) exerted its own pressure and that the total gas pressure was the sum of the individual pressures; water-

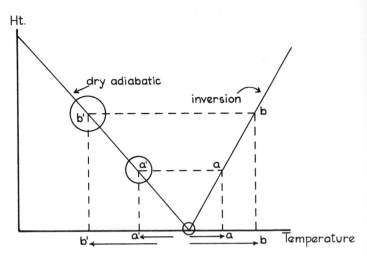

Fig. 18A The atmosphere in extremely stable equilibrium.

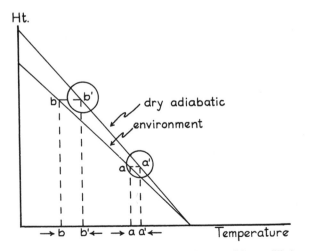

Fig. 18B The atmosphere in extremely unstable equilibrium.

vapor pressure is therefore simply that pressure exerted by the water molecules in the air present in the gaseous state.

The term *relative humidity* is the ratio of the actual water-vapor pressure to the maximum possible vapor pressure at the temperature, expressed as a percentage. Thus, when the relative humidity is 80%, the air contains 80% of the moisture it is capable of holding at that temperature.

The general gas law (p. 69) states that the total gas pressure is directly proportional both to the density and to the temperature of the gas. This implies that higher pressures are associated with higher temperatures.

At 100°C, the boiling point of water, the saturation vapor pressure is 760 mm of mercury or sea-level atmospheric pressure. That is, the pressure inside a bubble in a pot of boiling water has reached a value as large as the pressure of the atmosphere. Note particularly the lower left-hand portion of the curve in Fig. 19. Below 0°C the single curve divides into 2 curves, one referring to water-vapor pressure measured over a flat ice surface and the other to water-vapor pressure over a flat water surface. At the same temperature the saturation vapor pressure measured over water is slightly but significantly greater than that measured over ice.

The curve in the diagram shows that warm air can hold more water than cooler air, since pressure exerted by a gas is a function of both the number of molecules present and the temperature.

The percent of saturation, or the degree of humidification, can be increased by decreasing the temperature and/or by adding water vapor to the air, either through evaporation from a water source on the surface of the earth — the ocean, a lake, a river, a moist field, or transpiring plant and/or animal life — or through evaporation of water drops or ice crystals in the atmosphere or the exhaust of a jet engine.

As discussed earlier, the saturation vapor *pressure* can be reduced through cooling. This happens each night as the ground loses its accumulated solar energy back to space (or to the atmosphere) by long-wave radiation and the adjacent air experiences contact cooling with the cooling surface. However, by far the most significant method of reducing the saturation vapor pressure is through the dynamic cooling that accompanies ascent and expansion.

Saturation

If the humidification processes do not cease once the air has become saturated, the relative humidity should continue to increase beyond 100%. Were it not for the presence of the many fine particles in the air, this is what would actually happen. However, atmospheric processes rarely yield supersaturations greater than a few tenths of 1%. Even in the case of the updraft of a cumulonimbus cloud, where the air is moist and expanding very rapidly, the supersaturation does not exceed a few percentage points.

The condition of saturation exists when the concentration of molecules over a liquid surface is such that the number hitting and entering the surface is the same as the number leaving the surface.

If the surface is curved — the surface of a drop, for instance — the saturation vapor pressure over the surface is slightly increased, and the effect is further increased the greater the curvature of the surface.

Fig. 19 Saturated vapor pressure as a function of temperature.

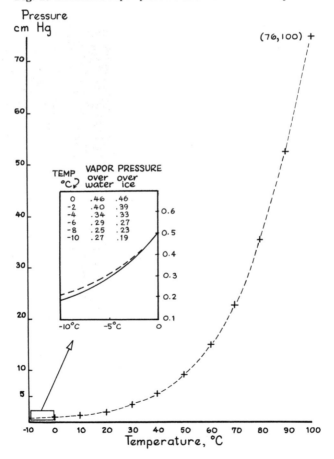

Condensation

When the relative humidity reaches 100%, nature provides an alternative that is preferable to a state of supersaturation. This is condensation of water molecules on those small particles in the atmosphere that are called cloud condensation nuclei (see Chapter 1). These nuclei start to gather water at relative humidities of less than or slightly greater than 100%. In general terms this means that these nuclei like to form chemical bonds with water that enable the water molecules to stick to the nucleus.

Condensation is a complex process. When water molecules in the gaseous state cluster about and stick to a nucleus, they lose their respective kinetic energies to the air; they release what is called *latent heat*. With water, this is a large amount of energy — nearly 600 kilocalories for each kilogram of condensed water, depending on the temperature and increasing as the temperature decreases. This release of latent heat is of great importance because it adds buoyancy to the air. It also changes the *dry adiabatic lapse rate* of $10°C/km$ to the *moist adiabatic lapse rate,* which may drop as low as $6°C/km$ depending on the rate of release of latent heat.

Sometimes condensation starts before the relative humidity reaches 100% if there is a population of hygroscopic — water-loving — nuclei. Giant (10-micron) salt nuclei produced by breaking waves along coastlines and over the vast ocean area are an example. They start to become active cloud condensation nuclei at humidities as low as 70%. This accounts for the typical haziness sometimes characteristic of coastlines; the haziness is a mist of fine water droplets formed on these nuclei.

Growth

Typically 30–1000 cloud condensation nuclei per cubic centimeter become activated in a slightly supersaturated environment. Following the activation, the growth of nuclei into droplets, and then drops, continues by diffusion of water molecules from the supersaturated environment to the drop surface.

The supersaturation required is provided by the cooling of ascending and expanding air. For cumulus (heap) clouds, this happens in the convective updraft. In the case of stratus (layer) clouds, particularly altostratus and cirrostratus, the expansion and cooling is due to a combination of slow upward motion and lowering pressures in the larger-scale weather systems of which the clouds are a part.

The overall dimensions of particle growth and relative volume are summarized in the table below.

TABLE 3

Typical Particle	Diameter microns	Diameter millimeters	Volume Relative to That of a Cloud Droplet
Cloud nucleus	0.12	0.00012	1/1,000,000
Cloud droplet	12	0.012	1
Large droplet	100	0.1	579/1
Mist	500	0.5	72,300/1
Drizzle drop	1200	1.2	1,000,000/1
Raindrop	3000	3.0	15,600,000/1
Heavy shower drop	6000	6.0	125,000,000/1

Precipitation

Clouds generate precipitation when the processes leading to growth have managed to increase the volume of a cloud droplet by a million times or more!

Two principal theories are used to explain how a cloud droplet grows to raindrop size.

The Wegener-Bergeron-Findeisen Theory As noted, water vapor pressure is slightly greater over water than over ice at the same temperature; this is of immense importance in cloud physics, and was first recognized by the German meteorologist A. L. Wegener in 1911. He pointed out that the coexistence of ice and supercooled water in a cloud would establish a varying field of vapor pressure between droplet and crystal, and he hypothesized that condensation would take place continually on the ice at the expense of the evaporating droplet until all the liquid water eventually disappeared. The process can be seen schematically in Fig. 20, where the greater saturation vapor pressure over a water surface (as compared to ice at the same temperature) causes supercooled droplets to evaporate and ice crystals to grow.

Another important contribution was made in the mid 1930s by the Swedish meteorologist Tor Bergeron through his classic paper, "On the Physics of Clouds and Precipitation." Bergeron incorporated Wegener's idea into a theory that suggested that every raindrop of diameter greater than 500 microns has its origin as a particle of ice. The implication of his theory was that the tops of a rain cloud must be colder than 0°C.

In order to appreciate what takes place in the top of a cloud that extends well above the freezing level, let us assume that the temperature in the cloud falls as low as −10°C, and that its top contains both ice crystals and supercooled water drops. A vapor-pressure difference exists between these drops and crystals, as shown in the figure. Water molecules diffuse toward the ice. As a consequence the water drops evaporate and the ice crystals grow.

At a temperature of $-15°C$ the preferred crystal growth pattern is the stellar form. As growth continues, a snow crystal is formed and if it becomes heavy enough, it may fall from the cloud, adhering to others as it falls. Depending upon various other factors — principally the thickness of the cloud and the altitude of the freezing level — the snowflake will either reach the ground as snow or melt enroute and become a raindrop.

Additional important contributions were made in the late 1930s by a German meteorologist, W. Findeisen, whose analyses of upper-air observations in northern Europe confirmed Bergeron's hypothesis. The Wegener-Bergeron-Findeisen theory served without serious challenge for a decade or so as a description of the primary precipitation mechanism, and still is of great importance.

Observations show the existence of a region from about $-12°C$ to $-17°C$ in which crystal growth rates are as much as 100 times greater than at warmer and colder temperatures. This zone of rapid growth occurs in the temperature region favoring stellar (dendritic) crystal growth. (Other temperature regions favor other crystal shapes — see Appendix 1.) In any cloud extending from $0°C$ to $-20°C$, crystal growth in the $-12°C$ to $-17°C$ region will overshadow that of all other levels. This helps explain the common observation that dendritic crystals are the most frequent crystal form for natural snow. With the combination of rapid growth rates and stellar crystal habit in the $-12°C$ to $-17°C$ temperature region, there is a marked difference in the precipitation behavior of

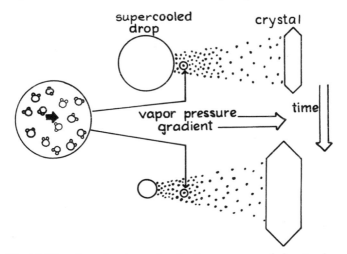

Fig. 20 Migration of water molecules from a supercooled water drop toward an ice crystal.

clouds depending upon whether or not they extend into this temperature range.

Coalescence Theory In the 1940s, new information showed that substantial quantities of rain did fall from warm clouds — clouds whose tops definitely did not reach the freezing level. This forced meteorologists to consider other mechanisms of raindrop growth, and growth by coalescence seemed most likely to explain the observations.

The very idea of collision suggests relative motion. There is no possibility of collision on a freeway when all the cars are moving parallel at the same speed in the same direction. In contrast, the most serious freeway collisions occur when one car is forced to stop and other cars in the traffic line successively pile into (and coalesce with) the obstacle ahead.

The motion of spheres of water should first be considered in relation to still air. The physical principle involved here is that an object accelerates until all the forces acting on it are balanced, and then it moves at a constant speed. A sphere of water in still air is acted upon by 3 forces: (1) gravity, (2) buoyancy of the air due to thermal differences, and (3) the frictional retarding force (the drag). The frictional force is proportional to the velocity. When the sphere falls at a speed at which the drag is equal to the difference between the force of gravity and the buoyant force, it has reached its terminal velocity for those conditions, as shown in Fig. 21.

Fig. 21 Distorted shape of an 8000 μm drop falling at its terminal velocity.

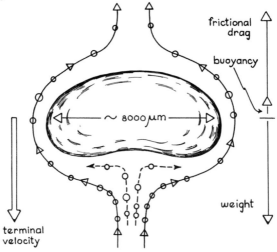

The terminal velocity for small cloud droplets is very low, on the order of millimeters per second (mm/sec.). Therefore it does not take much updraft in a cloud to keep cloud droplets suspended at the same level. The terminal velocity for raindrops is quite high by comparison — several meters per second (m/sec.).

Assume now a cloud in which the droplet size varies by 400 times, say from 20-micron to 8000-micron diameter, and all the droplets are falling through still air at their terminal velocity as seen in Fig. 21. An observer riding on the raindrop would see the small cloud droplets approaching at nearly the fall speed of the raindrop, because the cloud droplet's fall speed is so slow. The probability is very great that the small droplet would be swept around the larger drop in the aerodynamic airflow, as illustrated (just as when one drives along a highway most small insects are swept past the windshield; occasionally a larger insect doesn't make it, and splatters against the glass).

The mathematics of coalescence are such that as long as the ratio of diameter of droplet to larger drop is less than about 0.2:1, there will be no collisions and no coalescence. When the 2 drops are of about the same size, they rarely collide because their fall speeds are about the same. The highest probability for coalescence exists when the ratio of smaller to larger size is about 0.6:1, as would be the case if, in our first example, we had 2 drops of 600 and 1000 microns. Collision does not necessarily mean coalescence — some pairs of drops bounce on collision.

The two theories of raindrop formation are not mutually exclusive or contradictory. The W-B-F theory deals only with cold clouds in which supercooled water droplets and ice coexist, but coalescence takes place in such a cloud as well. On the other hand, when clouds do not reach the ice-stage temperature coalescence is the only way to explain the resultant precipitation, if any.

The many different kinds of precipitation are described in Chapter 9.

The remainder of this chapter consists of photographs of clouds that commonly occur in the global atmosphere. While most of them were obtained by the authors in various parts of the U.S., a few are from other regions of the world. These photos illustrate the fascinating variations of cloud types. These cloud types reflect the properties of the atmosphere where they are formed, and provide much information to the observer. The types and quantities of cloud nuclei, the vertical moisture profile, the wind patterns, the air quality and stability, and other features can often be inferred from watching them.

Pl.41 I$_L$ Fair-weather cumulus over Grand Canyon. Note very flat bases and rounded tops. Clouds in background are merging into strato-cumulus (layered heaps). This is a settled-weather cloud.

80

Pl.42 I$_L$ Fair-weather cumulus tending toward stratocumulus in re-gions of most active development. Cloud over hill has extremely flat base and rounded top characterizing this settled-weather type.

Pl.43 I_L Fair-weather cumulus clouds have flat bases. These mark altitude at which rising moist air reaches its dew point. The convection plume producing these clouds rises from the earth.

Pl.44 $I_L III_M$ Fractocumulus clouds result from dissipating fair-weather cumulus as cloud droplets evaporate, due to their mixing with dry air, strong dry wind, or other diluting processes.

Pl.45 IV$_L$ Unusual form of convective cumulus, forming in moist sea air that is being forced up the mountain slope heated by the sun. The unstable air rises rapidly in narrow columnar towers.

Pl.46 IV$_{LM}$ Stratocumulus clouds at edge of the Japan Sea where heavy snowstorms occur. In the distance are cumulus that have been intensified as moist sea air rises due to mountains.

Pl.47 I$_{LM}$ Vigorous swelling cumulus with fragments of fractocumulus in foreground. Note large number of individual cells making up main cloud mass. These clouds will soon become cumulonimbus.

83

Pl.48 I$_{LM}$ Telephoto view of tops of swelling cumulus. Towers show presence of updrafts, each column being comprised of many small convective cells. Cloud-free space between towers in downdraft.

Pl.49 I$_{LM}$ Swelling cumulus with crepuscular rays that result from the greater amount of light-scattering in the direct beams of the sun. Darkness of cloud due to shadowing effect but note bright edges.

84

Pl.50 I$_{LM}$ and **I$_{LH}$** Swelling cumulus, and cumulonimbus in the rear where ice-crystal anvil is seen. Central swelling tower soon will be transformed to ice and become an active cumulonimbus.

Pl.51 I$_{LH}$ Exceedingly vigorous swelling cumulus about to turn into a cumulonimbus. Note sharp cauliflower profile and large number of convective cells present. Dark clouds are called scud.

85

Pl.52 I$_{LH}$ Nonprecipitating cumulonimbus sitting over San Francisco peaks. Base of similar cloud in upper foregound. Ragged base indicates imminence of precipitation. Cloud tops very high.

Pl.53 IV$_{LM}$ Large cumulus cloud intensified by the mountain. Partially obscuring the top of distant cumulus a mass of ice crystals falls from a thunderstorm moving toward the mountain.

86

Pl.54 IV$_H$ Cumulus tower that has risen into a region of moist air and caused the lifted moist air to cool below its dew point. The cloud formed in this manner is called a pileus.

Pl.55 I$_{LH}$ Young cumulonimbus in background with tops at about the ice crystal stage. Substantial amounts of wispy fractocumulus remain from dissipating cumulus in regions of gently sinking air.

87

Pl.56 I$_{LH}$I$_{LM}$IV$_H$ Evening sky with giant cumulonimbus obscuring sun. Crepuscular rays are seen below and at left. "Silver lining" due to light-scattering. Swelling cumulus left, cirrus at top.

Pl.57 II$_L$ Thin stratus lying in the valley and topping Pikes Peak. Ground fog in upper valley is starting to burn off, yet lower valley remains obscured. Fog is stratus on the ground.

Pl.58 II$_L$ Stratus cloud (ground fog) fills the shallow valley. Notice that the fog is starting to become wispy and uneven in form as it experiences the warming effects of the early morning sun.

Pl.59 II$_M$ Altostratus with a few stratocumulus in background. Ground-glass effect from ice crystals is shown in upper center as sun's rays only partly penetrate the gray cloud layer.

Pl.60 III$_M$ Altocumulus, rather high, aligned in large left-to-right rows or billows. A few low stratocumulus over hills at lower left. The higher clouds are typical formations in a jet stream.

Pl.61 III$_L$ Stratocumulus formed by spreading of cumulus whose tops are still somewhat in evidence, as at middle left. A few isolated cumulus of fair weather can be seen with some fractocumulus interspersed.

90

Pl.62 III$_L$ Stratocumulus, rather high, with some evidence of virga from elements at left center. A good example of the layered heap cloud, vertical motion limited by inversion.

Pl.63 III$_L$ Stratocumulus in an extensive pattern over prairie land; a classical example of the common lower layered heap cloud type. A thick-thin appearance indicates presence of up-down motion.

Pl.64 III$_L$ Stratocumulus in patches. Base of this cloud layer is about 1 km (3300 ft.) high and thickness is about 300 m (1000 ft.). This type of cloud results when free convection is limited.

Pl.65 III$_M$I$_L$ Altocumulus in several layers along with small patches of much lower cumulus. The lower patch of altocumulus exhibits small waves while the higher ones are random convective cells.

92

Pl.66 III$_M$ Low altocumulus covering whole sky. Backlighting dramatizes globular structure characteristic of altocumulus. Cloud elements have some alignment in rows in addition to cell activity.

Pl.67 III$_M$ Altocumulus in classical pattern. The predominance of cloud elements over clear spaces suggests general ascent in the air within which the altocumulus instability is found.

Pl.68 III$_M$ Altocumulus exhibiting much activity in the many convection cells, particularly in central and lower right areas of photograph. At upper right there is evidence of billow structure.

Pl.69 III$_M$ Altocumulus; rather thick-looking sections in lower rear are due to perspective. A very typical altocumulus sky, representing a condition of limited instability at that level.

94

Pl.70 III$_M$ Altocumulus nearly directly overhead, base about 5 km (16,500 ft.) wide and perhaps 0.5 km (1650 ft.) thick. Ascending and sinking air very well marked by cloud elements and clear space.

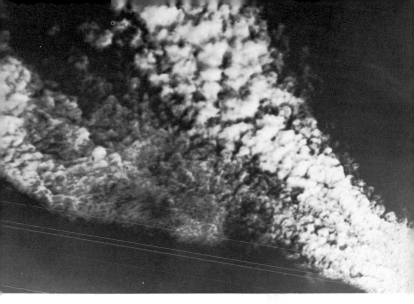

Pl.71 III$_M$ Altocumulus running across sky in a ribbon. Convection shown by white cloud puffs. Compensating downward motion noted by dark intervening regions. These are jet stream clouds.

95

Pl.72 III$_{M,H}$ Unusual altocumulus sky showing various-sized elements of convective activity. Individual cells in upper right are 10–20 times the size of those in lower right. Sky is changing rapidly.

Pl. 73 III$_H$ Cirrocumulus sky with unusual shadow effects cast by parts of lower cloud between observer and sun. In upper right, cloud is aligned in rows or rolls. Some cirrus is mixed in.

96

Pl. 74 III$_H$IV$_H$ Cirrocumulus, mainly at right, and cirrus, left and lower left. Cirrus apparently lies somewhat lower than the cirrocumulus. Note patterns of row alignment in the latter.

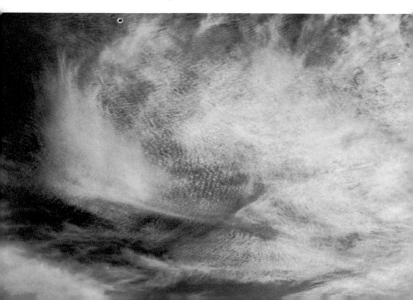

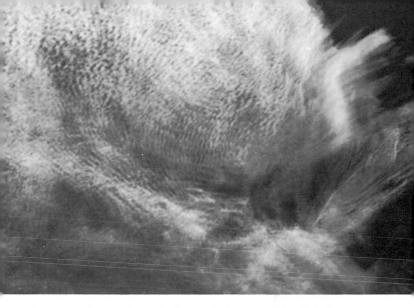

Pl.75 III$_H$ Cirrocumulus with cirrus principally at right. Globular pattern shows organization into rows or billows. The supercooled water droplets of cirrocumulus are changed to ice of cirrus.

97

Pl.76 III$_H$IV$_H$ Cirrocumulus, mainly at right, and cirrus, at left, in an intriguing pattern. Note organized pattern of waves in center of photograph somewhat obscured by lower cirrus.

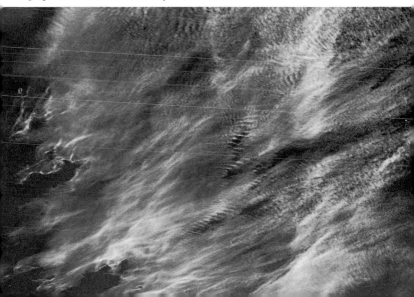

Pl.77 III$_H$ and IV$_H$ Cirrocumulus and cirrus at left. These types often appear together. Cirrocumulus appears in rolls as at lower right and center. General transformation to cirrus at upper right.

98

Pl.78 III$_M$IV$_H$ Chaotic sky consisting of altocumulus layers, higher at upper right, which shows presence of billow wave patterns. Dark cloud in center is much lower. Cirrus in lower rear.

Pl.79 III$_{LM}$ Very dramatic example of convective column spreading out as a stable layer is reached. Second tower of swelling cumulus penetrates first stable layer, then spreads out at second one.

99

Pl.80 III$_{LM}$ Stratocumulus cloud deck with isolated cumulus at center of photograph attempting unsuccessfully to penetrate upper inversion and forced to spread out, merging with the stratocumulus.

Pl.81 III$_{LM}$ Swelling cumulus spreading out laterally and forming layers of stratocumulus, the upper resembling low altocumulus. This is an excellent example of mixed heap and layer clouds.

Pl.82 IV$_{LH}$ Cumulonimbus, massive. Strong stable layer limits further vertical development; cloud spreads laterally producing collar or roll of ice crystals. Lower clouds represent many cumulus types.

Pl.83 III$_L$ Altocumulus, low, showing lessening convective activity. Stringy edges of cells seem to be losing their organization. In an hour or so this cloud will have disintegrated.

Pl.84 III$_M$ Altocumulus, low, with globular elements aligned in long rows similar to cumulus "cloud streets." Convergence of rows is a result of perspective and is an illusion.

Pl.85 IV$_L$ Nimbostratus. Dark and threatening cloud is generating a shower in center rear of photograph. This scene shows the underside of a cumulonimbus cloud mass whose tops are obscured.

Pl.86 IV$_L$ Nimbostratus storm cloud and widespread precipitation falling through a very ragged lower cloud, separated elements of which are called scud.

Pl.87 IV$_L$ Nimbostratus with precipitation obscuring cloud mass from which it falls. Fractostratus at upper left and swelling cumulus towers visible through precipitation at lower right.

Pl.88 IV$_L$ Nimbostratus and cumulus in all stages of development after passage of active cold front. The presence of copious amounts of scud accentuates stormy appearance of sky. Rain has just ended.

Pl.89 IV$_M$ Nimbostratus cloud system generating widespread rain that is heavy enough to reduce visibility. If the observer sees such precipitation develop from a cumulus buildup, the group is IV$_{MH}$.

104

Pl.90 IV$_{LH}$ Cumulonimbus showing a high flat base from which a heavy rain shower falls. Note many convective towers in cumulus cloud mass with cumulonimbus anvil at right rear.

Pl.91 IV$_{\mathrm{LH}}$ Cumulonimbus in classic textbook dimensions. Note flattish base from which heavy showers are falling, the thousands of convective cells, and well-formed wispy ice-crystal top.

Pl.92 IV$_{\mathrm{LH}}$ Cumulonimbus. This massive cloud is particularly interesting in terms of the double ice-crystal collar, suggesting a weak lower inversion that was penetrated, and strong lid at stratosphere.

Pl.93 IV$_{LH}$ Ocean rain squall from cumulonimbus. Dark mass in center of photograph is precipitation falling from base of cloud mass. Sun is shining in background, indicating this is an isolated cloud.

Pl.94 IV$_{LH}$ Savage-looking sky with very heavy rain falling right of mountains from an extremely vigorous cumulonimbus system. Under proper conditions this kind of cloud could produce a tornado.

Pl.95 IV$_H$ Cirrus covering most of the sky in no particularly organized pattern. Thickest areas at upper left and lower right mark regions of heaviest ice-crystal precipitation.

Pl.96 IV$_H$ Cirrus showing centers in which ice crystals are generated. The streamers of falling ice crystals tend to lag behind at right, gradually dissipating in the warmer air below.

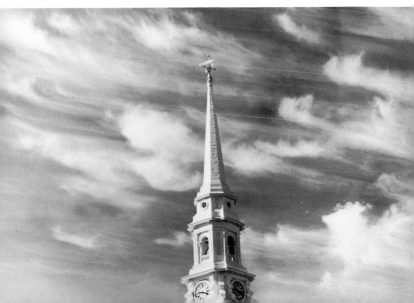

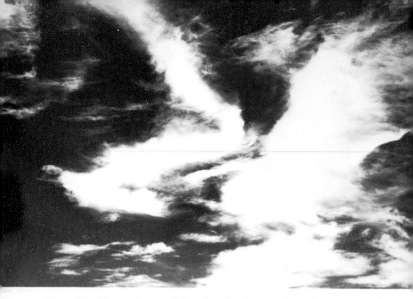

Pl.97 IV$_H$ Cirrus, dense, of the type that is sometimes the remnant of cumulonimbus anvils. Central cloud elements are generating heavy concentrations of falling ice crystals.

Pl.98 IV$_H$ Cirrus, dense, exhibiting active plume of falling ice crystals particularly in center of photograph. Cloudless regions represent dry areas between tongues of moisture.

Pl.99 IV$_H$ Example of fibrous structure of cirrus. Seen from a high-flying plane this cloud may be several km thick. Appearance results from many trails of falling crystals in perspective.

Pl.100 IV$_H$ Tufts of dense cirrus seen directly overhead. Such a cloud is produced when air at this level experiences uplift and cooling, with resulting ice crystals growing heavy enough to precipitate.

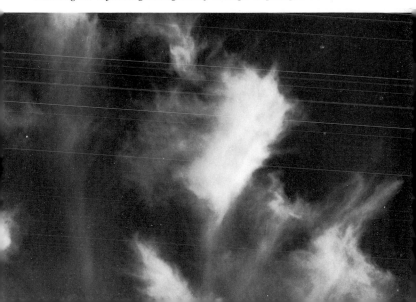

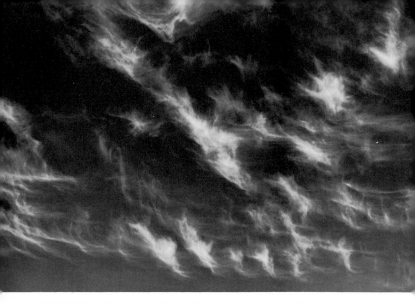

Pl.101 IV$_H$ Cirrus in a delicate fibrous pattern caused by streamers of falling ice crystals that dissipate in the warmer lower regions. It is likely that these clouds are in a jet stream.

Pl.102 IV$_H$ Cirrus in a jet stream having large differences in velocity. The upper portion of the fibrous clouds of ice crystals has a much higher speed than the long trailing streamers below.

Pl.103 I$_L$ and II$_H$ Classic example of fair-weather cumulus with a cirrus veil at upper right. Cumulus in many stages of growth, cloud bases about 1 km (3000 ft.); thickness about 0.5–1.5 km (600–4900 ft.).

111

Pl.104 III$_{LM}$IV$_{LH}$ Fair-weather cumulus mixed with low altocumulus and a large cumulonimbus at left rear. Lifting condensation level is same, producing uniform base of all the cumulus type clouds.

Pl.105 $I_{LM}I_{LH}IV_H$ Cumulus in various stages of development with highest tower starting to spread laterally and turn to ice. Substantial amounts of fractocumulus are present along with thin cirrus.

112

Pl.106 $I_{LM}II_H$ Mixed sky of cumulus in several stages of development and layer of cirrostratus seen at top of photograph. Note the two well-organized swelling towers center and upper right.

Pl.107 III$_{LM}$IV$_H$ Mixed sky consisting of active cumulus reaching stable layer and spreading out. Precipitation is falling from false cirrus in center of photograph. Middle layers seen at lower left.

Pl.108 I$_L$IV$_H$III$_{LH}$ Mixed sky consisting of cumulus masses with an anvil being blown from right to left. Isolated small cumulus in upper part. Cirrus in top half of photograph. Swelling cumulus lower left.

Pl.109 III$_{LH}$IV$_H$ Unusual layering effect showing remnant of cumulo-nimbus tops (anvils) generating streams of ice crystal precipitation. A strong left-to-right wind at cirrus level can be inferred.

114

Pl.110 V (mammatiform) and I$_{LM}$ Swelling cumulus. Upper portion of photograph shows active pouches or mamma developing on under-side of anvil of cumulonimbus whose cirrus top extends from right to left.

Pl.111 IV$_{LH}$ Cumulus system in all stages of development. At lower rear heavy showers are falling from cumulonimbus whose top is not visible. Lightning and thunder can be expected in this storm.

115

Pl.112 IV$_{LH}$ Dark and threatening underside of a matured cumulonimbus from which precipitation is falling at lower left. Center small cloud is called scud. Thick stratocumulus hangs over mountains.

Pl.113 III$_{LM}$IV$_H$ Mixed sky made up of cumulus, strato- and alto-cumulus. Towers of swelling cumulus rise in left background through broken lower layers. Dense cirrus upper right is precipitating.

Pl.114 I$_{LH}$I$_{LH}$III$_L$ Very large cumulonimbus at left is brightly lighted. Active swelling cumulus in foreground is in early stages of cumulonimbus development. Layer of stratocumulus at lower left.

Pl.115 $I_L I_{MH} IV_H$ Cumulus, in different stages of development, and cirrus. Swelling cumulus, center, is growing vigorously aided by midday heating on mountain slopes facing observer. Cirrus center rear.

117

Pl.116 IV_H Jet stream cirrus above Rockies. This dramatic example is indicative of a strong jet stream. Although it appears thin the cirrus is perhaps 1–2 km thick. Note small cumulus over peaks.

Pl.117 III$_M$ Altocumulus with some fractocumulus in upper right of photograph. Note interesting patterns of convective cells in altocumulus layer. Thin areas correspond to sinking air.

Pl.118 II$_L$II$_H$ Stratus blanket covering valley east of Sierra Nevadas. Highway is seen disappearing under cloud deck that had ceiling of about 10–30 m (30–100 ft.). Upper cloud layer is cirrostratus.

Pl.119 IV$_L$III$_L$ Stormy stratocumulus mixed with nimbostratus. Photograph shows underside of large cumulonimbus cloud system. Glider in center is taking advantage of the vigorous updrafts.

Pl.120 I$_L$II$_H$ Cumulus forming over warm land while ocean is free of clouds because it is colder. Higher stratus mark the stable air that occurs in trade-wind areas. Typical clouds of moist tropical air.

Pl.121 III$_M$ Clouds forming at middle levels in the atmosphere at sunrise. Such clouds indicate the presence of large amounts of moisture likely to favor thunderstorm development.

120

Pl.122 III$_M$ A mass of altocumulus changing to ice crystals. The presence of such clouds strongly indicates that thunderstorms will develop later in the day.

Pl.123 IV_{MH} Shower starting to fall from the black base of a mature cumulonimbus. Note bits of scud beneath base. Intensity of rain will probably increase with time; thunderstorm could result.

Pl.124 IV_{MH} Towering cumulus consisting of a large number of convective columns that have merged into a single storm. The top of the cloud will soon spread out to form an anvil.

Pl.125 I$_{LM}$ Early cloud formation from moist air rising above sun-heated mountains, commonly begins two hours after sunrise. Start of the thunder–hail storm seen in Plates 126–132.

Pl.126 I$_{LM}$ Within 20 minutes the clouds begin to merge into a single system. Top and sides of developing storm resemble a cauliflower. Base looks dark because it receives little light.

Pl.127 I$_{LMH}$ After about 90 minutes the top of the towers begin to look fuzzy, due to the development of ice crystals. At the same time shafts of precipitation appear at the cloud base.

Pl.128 I$_{LMH}$ The storm has reached a new phase in which towering plumes emerge from central region of cloud mass. A series of towers appear in succession, each larger than the previous one.

Pl.129 IV$_{LMH}$ As the storm develops the tops of the towers subside, change to ice crystals, and move beyond the mountain. A white veil of precipitation appears at the bases of the older clouds.

124

Pl.130 IV$_{LM}$ In less than two hours from the formation of the first tiny cumulus clouds above the mountain, heavy rain or a mixture of rain and hail reaches the ground. Lightning often occurs.

Pl.131 IV$_{LM}$ The heavy precipitation has moved beyond the mountain and is likely to continue for another hour or two. Adjacent storms may move into the region accompanied by more lightning.

125

Pl.132 IV$_{LM}$ Summits of mountain are white with hail. Most often hail occurs at beginning of heaviest precipitation and most intense lightning. A general overcast remains until evening.

Pl.133 V Parallel lines of clouds called cloud streets, which often extend for many miles in air that is free of horizontal discontinuities.

126

Pl.134 III_L A cloud street produced in the lee of a large mountain. Stable moist air displaced by mountain barrier converges to form a line of clouds in what is termed a "wake effect."

Pl.135 I$_M$III$_M$ Moist air displaced upward in unstable air forms a complex mixture of wavelike clouds with cumulus towers extending above. A good example of mountain effect in cloud development.

Pl.136 I$_M$III$_M$ An excellent example of what happens when moist, partially stable air is forced upward over a barrier, forming unique clouds on the mountain.

Pl.137 II$_M$III$_M$ Mixture of altostratus and some bands of altocumulus in center and lower portions. Sun shines through altostratus as if through ground glass. Central band dark because of shadowing.

Pl.138 I$_M$II$_H$ Two different cloud types are present. The lower cumulus are formed over mountains while the much higher stratiform cirrus consist of ice crystals as indicated by the fuzzy sun.

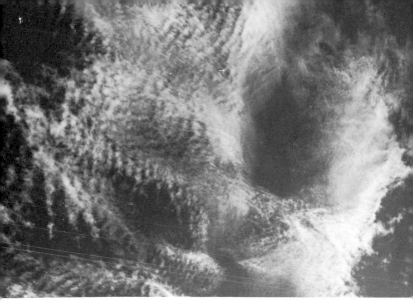

Pl.139 III$_H$ Cirrocumulus forming in a region having considerable shear. This produces ripples in the cirrus sheets, with the supercooled droplets being converted to ice crystals.

Pl.140 III$_H$IV$_H$ Clouds consisting mostly of cirrus drawn into long filaments and streamers, some of which have developed ripples. Several isolated cells of supercooled clouds are present.

Pl.141 IV$_H$ Streamers of ice crystals formed in supercooled cirro-cumulus; the crystals are falling into air that is moving at a much higher velocity, causing the streamers to develop.

Pl.142 IV$_H$ Ice crystals falling in long streamers from generating cells that consist of supercooled water droplets. In this instance there is fairly uniform air velocity at cloud level.

Pl.143 IV$_H$ Cirrus covering the entire sky with long ice-crystal stream-ers trailing off to the left (west). Note the central wispy elements in contrast to the thicker cirrus regions in the background.

131

Pl.144 Cirrus clouds near the base of the stratosphere show evidence of the shear effects that occur when fast-moving air comes into contact with a stable layer of air.

Pl.145 V Condensation trail from a jet plane that produces its shadow on a diffuse mass of ice crystals extending downward below the contrail and the sun.

132

Pl.146 The condensation trail of a jet plane that is growing where moist air is supersaturated with respect to ice. This shows that the contrail is located below the thin streamers of cirrus cloud.

Pl.147 The effect of the sun's heat on a hillside covered with a thin layer of snow. As the snow melts a cloud forms and moves up the heated slope into the forest.

Pl.148 A long stream of altocumulus billow clouds marks the under-surface of a jet stream. The clouds move rapidly with the long axis of the billows at right angles to the flow of air.

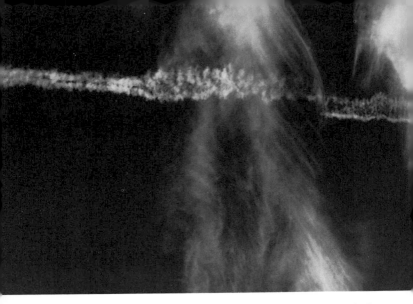

Pl.149 A shear zone near the base of the stratosphere. The jet airplane condensation trail is displaced as well as changed by the increased turbulence and moisture below the cirrus cloud.

Pl.150 Another example of the differences in wind-shear velocity and moisture that often exists near the base of the stratosphere. Long streamers of ice crystals flowing from the contrail result.

3

Unusual Clouds in the Atmosphere

It is hoped that this guide will encourage the reader to glance at the sky more frequently. Most of the clouds that are seen will fit into the categories described. Occasionally, however, the ordinary weather will be displaced by unusual atmospheric events. The sky will be different under these relatively rare conditions; unusual clouds are likely, and for interested photographers, this is the time to try to record extraordinary atmospheric effects.

Several predictable conditions are likely to produce unusual clouds. Perhaps the most frequent of these occurs when a jet stream is overhead or swinging by. In any part of the U.S. or any other place above 20° north or south latitudes, jet stream clouds may be visible. Some areas, such as the northwestern and northeastern parts of the U.S., are favored, but wherever intensive weather systems pass, chances are good that a jet stream will be located within the edge of the high pressure cell. Its location tends to be at the boundary between warm and cold air masses — a cold front near the stratosphere.

Thus, at ground level the observer is likely to be in the cold, clear air following a frontal passage; this is good for observing as well as photographing the phenomenon. Since the jet stream is a river of air coursing through the atmosphere at velocities often above 60 m/sec. (130 mi./hr.), it is quite likely that the clouds will be spectacular, if there is sufficient moisture in the air to form them. This is especially true in mountainous country or when the jet stream encounters a mass of dense air that behaves like a mountain barrier and causes the fast-moving air to produce waves or billows. The waves that form in this manner are often quite beautiful. Standing wave clouds (see Plate 165) are generally caused by certain mountain configurations. Under some conditions the wave pattern may form at a considerable distance from the mountain that produces it. This is determined by the velocity of the wind in the jet stream and its effect on the lower air, which it may drag along by a kind of aspiration.

Traveling wave clouds in a jet stream have several probable causes. Just as with the traveling wave in a sand-laden stream that is produced when the density of the suspended load of sand grains becomes high enough to act as a barrier, this type of wave in the atmosphere may be initiated by a complex interaction of stable

135

and unstable regions of the air at the jet stream interface. In other instances a wave cloud may be induced by a small hill that is not large enough to continuously control the position of the undulating air being driven by the jet stream.

Because of this dynamism and because the jet stream is a ribbon of air rather than a thick mass, the visible effects are generally quite transitory and the most spectacular effects last only a short time.

Another high, evanescent cloud form is the pileus. Often described as a zone of ice crystals that appears on top of, but separated from, a towering cumulus, this cloud is produced by the upward displacement of air that contains enough moisture so that the upward movement brings the moisture in the air to saturation. If this cloud forms at a temperature colder than $-40°C$ ($-40°F$) — the Schaefer point — it will consist of ice crystals. However, clouds of this kind often form at temperatures of $0°C$ to $-30°C$ ($+32°F$ to $-22°F$) and thus are often supercooled. While this cloud is most commonly seen above the top of large cumulus towers, it also commonly forms at the top of standing waves. At times this type of formation may resemble a huge stack of flapjacks.

Another fascinating and spectacular cloud form is the mammatiform — clouds often seen falling as rounded protuberances from the undersurface of a giant anvil top spreading from a cumulonimbus cloud. These tend to occur whenever "cloudy air" comes into contact with a stable layer of cloud-free air. In a short time the cloudy air reaches the same temperature as the cloud-free air because of radiation and mixing. Since some of the cloudy air above is now at the same temperature, it is unstable because of the aggregate weight of the cloud droplets. However, because it has no nearby boundaries, the falling air divides into a polygonal array of downward-moving parcels of air called Bénard Cells. Since such downward movement must have upward compensation of cloud-free air, the undersurface of the cloud soon becomes a mass of pendules which, as they fall, assume the rounded shapes that become a spectacular sight, especially when illuminated by the setting sun.

A rare and seldom seen but fascinating cloud form is known as the noctilucent cloud. Generally seen after sunset or before sunrise at high latitudes (they are most often seen in the Scandinavian countries, northern Canada, and similar sub-Polar regions), it is a jet-stream-like wavelet cloud whose appearance suggests that it has formed in very high velocity air. These clouds form in the upper stratosphere at elevations of 80 km (50 mi.). At this altitude the atmosphere has practically no water vapor so the constitution of those clouds is a question now being studied. Because of their evanescent beauty, the authors hope the reader will be privileged to see them, whether on a tour ship, an airplane — on a polar flight, or as one of the hardy souls who lives in these latitudes.

There are of course many other forms of unusual clouds. Giant convective clouds such as thunderstorms, hailstorms, tornado clouds, hurricanes, and many other organized systems, present an endless series of challenging opportunities to see, describe, or better yet photograph atmospheric phenomena. Generally the observer must be on the periphery of such clouds to have the light and the proper perspective to see them. In time, for those who become interested, experience will show that it is unwise ever to be far away from a camera!

In addition to the unusual clouds that appear in the "natural" atmosphere there are others that occur because of some sort of cataclysmic or unnatural mechanism. Examples are the clouds formed by the heat above smokestacks condensing the atmospheric moisture on a cold morning, and the clouds that form above electric generator cooling towers wherein the moisture comes from the aeration of hot water.

Nature also produces some very unusual clouds in sporadic instances. These include mixtures of water droplets, gaseous vapors, and ashy residues blown out of the earth by volcanoes, and the spectacular water droplet clouds that are emitted by geysers in places like Yellowstone Park.

In this era of high-flying aircraft, condensation trails — contrails — are a commonly observed feature of the sky. Sometimes they are ephemeral and dissipate as quickly as they form; other times they persist and grow wide enough to cover a substantial portion of the sky with a sheet of cirrostratus. Sometimes they maintain their initial integrity as a line of cloud formed in the wake of the rapidly moving aircraft; at other times they develop a series of pendules from which streamers of precipitation are observed to fall.

Contrails are a fascinating subject for study, sufficiently complex to challenge the expert and sufficiently variable to intrigue the amateur observer. Properly understood they yield a wealth of information about the current state of affairs in the high atmosphere, where it is difficult to locate weather instruments. Observed systematically, as a function of time, contrail information is a valuable adjunct to forecasting the weather.

Chapter 2 explained that a cloud forms when the moisture content of the air at a particular temperature exceeds a critical amount and that this condition can be attained in 2 ways: (1) adding water vapor to the air from an external source, or (2) cooling the air, thus reducing the amount of moisture it can hold. A second important fact that was discussed (and is illustrated in Appendix 16) is that at a given temperature, slightly more water vapor can be held over a water surface than over an ice surface.

It is easiest to consider the contrails laid down by commercial jets that commonly fly from 10 km to 13 km (32,800–42,600 ft.)

where the temperatures range from $-30°C$ to $-65°C$ ($-22°F$ to $-85°F$). This is the region of the high troposphere or the low stratosphere. It takes very little water at these temperatures to produce a condition of saturation or supersaturation. Between these extremes lie all the possibilities for the many variations in pattern that occur. Moisture at high elevations often advances through the sky in tongues or uneven patterns, both large and small. Thus contrails may be seen in uneven segments of growth and dissipation.

Sometimes the multiple contrails that persist will break into a series of swirled loops, with the loops joining into ovals. When these loops and connected vortices develop, they ordinarily do not last more than a minute or so. When trails last for longer periods, they may break into a series of pendules or fingers. The pendule is a form of ring vortex, indicating the presence of locally stable moist air. Sometimes pendules are pulled away into long streamers by a zone of faster moving air. The resulting shear motion may result in such an extensive cloud sheet that the sky becomes completely overcast.

Contrails are spectacular in early morning and evening, and even in the light of the full moon. When the air to the west is clear and the sun has set, a jet aircraft at 12 km (about 40,000 ft.) will still be illuminated by the sun. Sometimes the white contrail will assume a brilliant red or orange color. To the uninformed, the moving trail, produced by a plane that is invisible or appears only as a rapidly moving spot in the sky, may seem to be an out-of-this-world apparition — more than one such case has subsequently been headlined in the papers as a UFO with a fiery tail.

Under most conditions a contrail is made of ice crystals. Though it initially consists of liquid water droplets in the warm exhaust air, these soon change to ice particles in the frigid temperatures of the high atmosphere. The presence of ice is best illustrated when the contrail is located between the observer and the sun. Bright areas called parhelia (see Color Plate 11) are visible at an angle of $22°$ on either side of the sun. This angle can easily be checked with a built-in device possessed by every human. With arm and fingers outstretched, let the thumb obscure the sun. The little finger, stretched to the maximum, then subtends an angle of about $22°$. The bright spots observed are a portion of the well-known $22°$ halo (see p. 158).

The jet aircraft disturbs its environment in 2 important ways. First, a jet engine consumes large quantities of fuel, and substantial amounts of water vapor, a major by-product of combustion, leave the engine as one component of the exhaust gases. Second, the rapid movement of the air over the wings and body of the aircraft generates vortices that persist for a time until their internal energy is dissipated.

The particular contrail pattern will be determined by several interacting factors. These are: (1) the moisture content of the air, (2) the temperature of the air, (3) the moisture introduced into the air by the engines, (4) the vertical stability of the air below, at, and above the aircraft, (5) the cloud condensation and ice nuclei count when the air is warmer than $-40°C$ ($-40°F$), and (6) the mixing that takes place between the environmental air and the exhaust.

At one extreme, the air through which the jet is flying might be gently sinking over a large area, and thus have a very low relative humidity. In this case, the addition of moisture might be insufficient to produce anything but a very short-lived contrail, quickly destroyed by mixing with the dry surrounding air. When a jet passes and leaves no contrails in the sky, or contrails that quickly disappear, this is a good prognosticator of fair weather.

At the other extreme, the jet flies through air that is gradually rising and becoming cool, and so may be nearly saturated. The addition of moisture from the jet exhaust may then be enough to produce saturation and consequent water drops or ice crystals. A persistent trail is an indicator of moist air, which may be the first sign of an extensive storm area moving into the region. This is particularly true when cirrus clouds are also present.

Pl.151 IV$_H$ Classic example of jet stream cirrus covering whole sky. Strong jet stream winds in upper portion of photograph leave behind nearly horizontal streams of falling ice crystals.

Pl.152 IV$_H$ Jet stream clouds. These wispy ice-crystal cirrus leaving long streamers behind are embedded in the fast-moving air of a jet stream. Velocities of 60 m/sec. (120 mi./hr.) are common.

140

Pl.153 IV$_H$ Jet stream clouds. This cloud is a mixture of water droplets and ice crystals. The ripples and tiny cells are cirrocumulus while the wispy regions are ice-crystal cirrus.

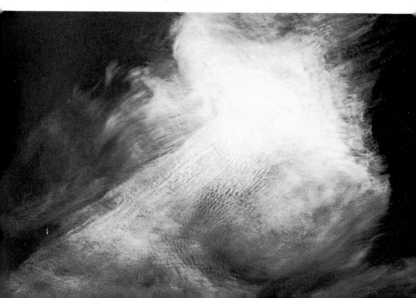

Pl.154 IV$_H$ Jet stream clouds. Great sheets of cirrocumulus often mark the upper levels of a jet stream. These clouds consist of liquid water droplets even though the temperature is close to $-40°C$ ($-40°F$).

Pl.155 IV$_H$ Jet stream clouds. The upper region consists of water while the lower, smoother clouds are ice crystals. Such clouds often extend for 1000 km or more and move and change rapidly.

Pl.156 IV$_H$ Jet stream clouds. A portion of the edge of an extensive line of cirrocumulus clouds that are changing from supercooled droplets to ice crystals. Central portion consists of crystals.

Pl.157 III$_M$ Jet stream clouds. Altocumulus lenticularis of the standing-wave type. These lens-shaped clouds form above or downwind of a hill or mountain as strong winds produce waves in the air.

Pl.158 III$_M$ Jet stream clouds. The main mass of cloud is a traveling wave in the lower levels of a jet stream. Higher cirrus is in the upper parts of the jet stream. Traveling waves change rapidly.

143

Pl.159 III$_M$ Jet stream clouds. Altocumulus billows. Such clouds often mark the lower level of a jet stream. They are miniature waves very similar to those seen in water.

Pl.160 III$_M$ Jet stream clouds. A large sheet of altocumulus billows in a jet stream. Portions of the billows are changing from water droplets to ice crystals as indicated by the wispy streamers.

Pl.161 Very high velocity winds of a jet stream have carried localized masses of ice crystals into long streamers. Such clouds often are present when velocity exceeds 100 m/sec. (200 mph).

Pl.162 V Billow clouds in a jet stream that have formed near the base of the stratosphere. Temperatures colder than $-40°C$ ($-40°F$) transform liquid cloud droplets to ice crystals.

145

Pl.163 A varied assemblage of cirrus that does not show much evidence of the shearing effect commonly visible when a fast-moving jet stream is present.

Pl.164 Cross-bedded cirrus formed in a jet stream, which suggests the presence of a turning motion in the cloud system. This pattern is often seen when a jet stream dominates the clouds.

Pl.165 V A wave cloud induced by a mountain. The cloud below, which looks like a cumulus, is called a rotor cloud and marks the presence of extremely turbulent air. Such clouds are highly unstable.

Pl.166 A layer of fast-moving air has produced a series of wave clouds. Below it, in moist air, are stratocumulus whose vertical growth has been limited by warmer air where the wave clouds formed.

147

Pl.167 Wave clouds formed in the lee of a ridge with altocumulus in a rugged array in all directions. Such wave clouds may change appearance very rapidly and sometimes move downstream.

Pl.168 A mass of wave clouds dominate the sky. They have formed in a jet stream, and the clouds are so thick that they have greatly reduced the amount of sunlight reaching the ground.

Pl.169 A very unusual line of billow clouds that formed in a jet stream as a long narrow line of traveling waves and were generated by a small ridge of land.

Pl.170 Roll or billow clouds of a jet stream. Such clouds result from vertical motion combined with shearing forces. The ascending motion produces a cloud while the descending cloud evaporates.

Pl.171 Bulbous or mammatiform clouds forming at interface between cloudy and cloud-free air that is stable. This often happens below the spreading anvil of a large thunderstorm.

Pl.172 The thin veil-like clouds seen at the top of the cloud mass at the upper center are produced by vertical displacement of moist air and are called pileus.

Pl.173 Ice crystals falling from a condensation trail left by an airplane. Since most contrails form at temperatures colder than $-40°C$ ($-40°F$), their moisture quickly generates ice crystals.

Pl.174 Condensation trails made by jet aircraft — training or testing flight. Passenger planes rarely deviate in their course except in a holding pattern.

Pl.175 The awesome cloud that has been generated by the explosion of an atomic bomb. Smooth surface of cloud is caused by rapid displacement of moist air, which produces condensation.

Pl.176 A subsequent view of the violent turbulence in the atomic bomb cloud as seen from 50 miles. It rose 13 km in 120 seconds and spread to a diameter of 160 km in 12 minutes.

Pl.177 Highly stable air above the ocean. At times such clouds may rest on surface of water to form a persistent fog; at other times they may be several hundred meters above the sea.

Pl.178 A dense fog that has engulfed a mountain valley. It has spread into three small tributaries. Such a cloud might be quite stable since a layer of stratus cloud shields it from the sun.

Pl.179 The tops of a line of four towering altocumulus pushing through a sheet of more stable altostratus clouds. The tops are pulled away by a 250-km jet stream flowing over Montana.

Pl.180 Underside of the anvil of a giant convective storm. Its rapid expansion has produced a horizontal vortex near the leading edge. Such roll clouds often contain severe turbulence.

4

Color in the Atmosphere

The interpretations of the color plates in the center section consist of both descriptions and explanations (see p. 165). They are intended to answer questions such as: What causes the colors of the rainbow? Why is one sunset cloud orange and another pink? Why should cirrostratus clouds cause a ring around the sun? and so on. The explanations are necessarily limited in scope; if the reader wants more information, he or she may wish to go on to the optics section of a physics or meteorology text.

The blue sky is so commonplace that it is taken for granted but makes a good starting point.

The Blue of the Sky

The sky is less blue when the air is dusty, and the blue changes to white when tiny ice crystals grow in the high troposphere to produce a veil of cirrostratus. What, then, is the cause of that deep and saturated blue that is seen after a rainstorm has washed out the atmosphere? The particles that produce the color of the sky are the air molecules themselves.

The reason for this phenomenon is preferential light scattering, the explanation of which is attributed to Lord Rayleigh, eminent 19th-century British physicist. If there were no light scattering, the sky would be black except for the sun, stars, and illuminated planets. Astronauts have verified this, once beyond the earth's atmosphere.

Figs. 22 and 23 illustrate the general size dimensions of the problem. Fig. 22 shows particles of 3 diameters: 1 micron, 0.1 micron, and 0.01 micron. The wavelengths of red and blue light photons are shown in Fig. 23.

Fig. 24 shows how the size of particles in the atmosphere, compared to the wavelength of the light illuminating them, determines the intensity pattern of the scattered light. Rayleigh's Law indicates that the light energy scattered per unit volume of air containing particles smaller than 0.1 micron is inversely proportional to the 4th power of the wavelength of the illuminating radiation. To make this more specific, consider a volume of air that is simultaneously illuminated by a beam of red light and another of blue. Rayleigh's Law predicts that the intensity of the blue light scat-

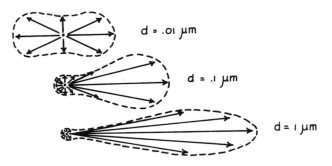

Fig. 22 Light scattering patterns of particles of three different sizes.

Fig. 23 Comparative wavelengths of blue and red light photons.

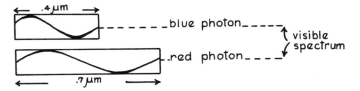

tered out of this volume will be 16 times the intensity of the red. An observer looking at this unit volume would sense an overpowering effect of energy in the short wavelength end of the visual spectrum and would describe its color by saying, "it's blue." So it is that the sky is blue.

The Colors of Clouds

If a cloud were a solid object with some known number of perfectly reflecting flat surfaces, it would be relatively easy to compute and predict its colors — assuming a clean atmosphere, knowledge of the sun's inclination, and the season; but the ideal situation is rarely approximated in nature.

A cloud is an aggregate of water droplets or ice crystals (sometimes mixed) of variable size. Cloud forms vary in shape. And the lower levels of the atmosphere always contain foreign particles, frequently in very large concentrations.

Allowing for these variables, it can be concluded that the degree of brightness and whiteness of the cloud depends on the relative position of cloud, sun, and observer, and upon the condition of the sky.

The cloud's illumination comes primarily from the sun, and secondarily from the scattered light it receives from the sky and other clouds. Dazzling white cumulus reflect white light directly from the sun to the observer. Another observer located on the opposite side of the cloud from the sun would likely describe it as dark gray.

Fig. 24 Rayleigh's Law—why the sky looks blue.

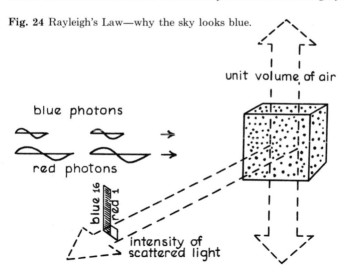

The exterior of a cloud is, of course, not a perfect reflecting surface. A whole complex of reflection, refraction, diffraction, and scattering mechanisms take place at the edge, and in the immediate interior of the cloud. The result is an outward flow of light from the cloud that the observer interprets as white, or as some shade of gray, or, at sunrise or sunset, as some color or combination of colors. Cloud bases are pink above desert sands.

The number of drops per cubic centimeter, the size of the drops, and the size *range* of the drops are important. These differences make it possible to see 2 cumulus clouds side by side, one brilliantly white and the other dark gray. The latter, by virtue of greater age, larger drops, and fewer drops per cubic centimeter is a more efficient absorber of light than the former.

Beyond the whites and grays of daytime clouds, there is the extremely challenging problem of explaining the magnificent array of cloud colors observed at sunrise and sunset, a small sample of which is displayed in the center section. Just before and after sunset and sunrise, the clouds receive illumination from the sun containing a greater proportion of red light (because of the scattering out from the beam of the blue light by the small particles in the atmosphere). A very great number of possibilities presents itself as we know from the varied shades seen in the morning and evening sky. Since the complexity involved in the explanation goes well beyond the limits of this guide, the interested reader is referred to such volumes as Minneart's *The Nature of Light and Colour in the Open Air,* which was a primary reference source for much of the material in this section.

Halo Phenomena

When cirriform clouds lie between the observer and the sun (or moon), refraction effects can lead to one or more of the numerous halo phenomena.

The most common — the 22°, or small, halo — occurs when the cloud form is a veil of cirrostratus and when the predominant ice crystal pattern is that of hexagonal prisms of uniform size. These are found at temperatures below approximately $-15°C$.

As shown in Fig. 25, a hexagonal prism consists essentially of two 60° triangles, base to base, with truncated tips. The angle of minimum bending from a straight line, called the angle of deviation, **D,** for a triangle with apex angle **A** $= 60°$ is 22°. This angle will be slightly greater for violet light than for red. As a consequence, we would expect to see a colored ring about the sun or moon with red on the inside, followed by yellow, green, white, and blue on the outside. The closest measurements for the inner edge give an angle of 21°51′ from the line between observer and the sun or moon. If the arm is outstretched, the 22° angle of the small halo is approximately subtended by the outspread hand measured from thumb to little finger.

The halo is caused by hexagonal columns that are tumbling in a random fashion, but only those oriented correctly at any given moment will produce the halo. The light rays that experience the minimum deviation will contribute most to the brightness of the halo.

The 46°, or large, halo is less frequently seen than its smaller counterpart, though its coloring is the same. As shown in Fig. 26, the minimum deviation angle of 46° results when the prism angle is 90°.

Circumscribed Halo When hexagonal crystals are rod-shaped rather than plate-shaped, and when the crystals oscillate about a horizontal axis, then larger arcs of light are seen. Usually these appear only as increases in brightness at the top and bottom of the small halo. Under unusually propitious circumstances, these are seen to be larger curves. The major types of halos are shown in Fig. 27.

Fig. 25 Refraction of light ray through hexagonal ice crystal that leads to halo phenomenon.

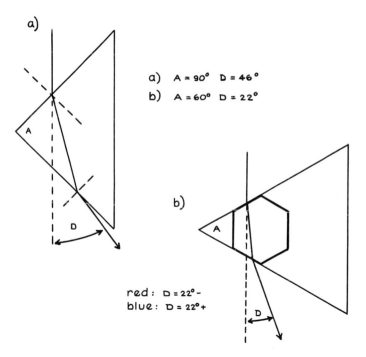

a) $A = 90°$ $D = 46°$

b) $A = 60°$ $D = 22°$

red: $D = 22°-$
blue: $D = 22°+$

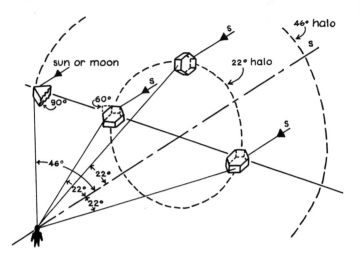

Fig. 26 Bending of light rays through hexagonal ice prisms produces a 22° halo.

Lowitz Arcs Vibrations in the tiny vertical ice crystals that produce parhelia ("mock suns") cause the Lowitz arcs. These arcs slope downward from the parhelia and touch the small halo. The phenomenon is rare and is seen only when the sun's altitude is high.

Light Pillar When the sun is low a column, or pillar, of light can sometimes be seen extending 5°–15° above it, and less frequently below it. The phenomenon is analogous to the pathlike patches of light formed by a rippled water surface early or late in the day.

In this case the pillar is supposedly due to reflections from ice plates rotating about horizontal axes and assuming different orientations in space; this explanation, however, is not well established.

Corona

A frequently seen, and very beautiful, phenomenon occurs when light from the moon shines through a thin layer of upper cloud. In its simplest form the corona consists of an aureole whose border, immediately next to the moon, is bluish, merging into yellowish white, with a brownish outer edge at a width of one or more moon diameters.

Quite frequently this aureole, or colored ring, is surrounded by other rings appearing in this order: blue, green (yellow), red; blue, green, red; blue, green, red. On rare occasions the third group can be seen outside the aureole; this is called the fourfold corona. The

diameter of the brown border of the aureole ranges from 2° to 10°.
(The diameter of the moon is about ½ an angular degree.) Coronae
can be seen around the sun, but with more difficulty because of the
sun's brightness.

The corona phenomenon is the result of diffraction of light
caused by the presence of small spherical water droplets in the
cloud. The purer the colors in the corona the more uniform the size
of the droplets. The larger the ring diameter, the smaller the cloud
droplets. Young clouds, such as waves and thin altostratus, are the
best corona producers since they have the most uniform droplets.
Older clouds are likely to have a greater variation in droplet size so
that the ring diameters overlap and weaken the color brightness.

Fig. 27 Major halo phenomena.

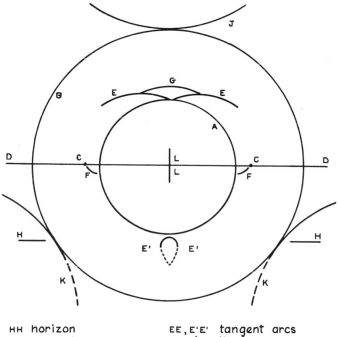

HH horizon	EE, E'E' tangent arcs
sun ~ 25° above HH	FF Lowitz arc
A 22° halo	G Parry arc
B 46° halo	J circumzenithal arc
CC two parhelia	KK infralateral arc
DD parhelic circle	LL sun pillar

These latter clouds have droplets whose diameters range from 10–20 microns (0.01–0.02 mm). Within a particular cloud, droplet size may be quite uniform.

Occasionally coronae are produced in cirrus clouds. When this happens the diffracting agent may be tiny ice crystals or frozen droplets of highly uniform size.

The Rainbow

When a single spherical water drop, as shown in Fig. 28, experiences an oncoming light wave front, most of the light passes on through the central portions of the drop, experiencing a lens focusing effect. This light plays no part in the formation of a rainbow; the "bundles of rays" that enter the extreme lower and upper portion of the drop do.

Fig. 28 shows a bundle approaching the upper edge of the drop; at **a** it is bent by refraction as it enters the drop. It strikes the interior surface at **b,** at an angle so great that the ray is reflected backward, strikes the drop's surface at **c,** and then emerges from the drop having deviated from its original direction by nearly 139°.

A second bundle that enters the lower edge of the drop, at **d,** experiences 2 internal reflections, at **e** and **f,** and is refracted at **g,** leaving the drop with a deviation of about 232° from the original

Fig. 28 Rainbow formation—shower drop bends light ray by refraction and internal reflection.

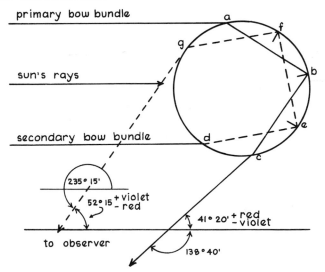

direction. This sets the stage for the primary and secondary bow.

The color sequence of the rainbow is also the result of refraction. Violet light is bent (slightly) more than red on passing through a prism. This means that violet light rays will deviate by a bit more than 138° and red rays a bit less. Therefore the vertical angle will be a little less than 41°20′ for the violet, and a little more for the red rays. As shown in Fig. 29, the observer will see in the primary bow an inner circle of violet light and an outer circle of red, with intermediate spectral colors between. In the secondary bow the colors will be reversed, but for the same reason.

Fig. 30 shows a sheet of raindrops illuminated by the sun's rays shining parallel to the earth's surface, as at sunset. Light arriving at the vertex point (the eye of the observer) will come only from a selected group of drops arrayed about the periphery of 2 cones, one whose vertical angle is 41°20′ and the other 52°15′.

A viewer standing with his back to the sun and holding an imaginary cone (such as a paper drinking cup) to his eye so that the extension of the axis of the cone intersects the sun, will be able to look along the surface of the cone and predict where a rainbow will appear in the sky. If the sun is setting or rising a full half-bow can be seen. Showers are more frequent in the late afternoon than in the early morning; rainbows are more likely to be seen in the afternoon for this reason. At noon no rainbows will be seen because the

Fig. 29 Primary and secondary bows of a rainbow.

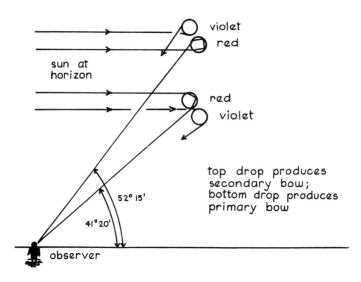

sun at horizon

violet
red

red
violet

top drop produces secondary bow; bottom drop produces primary bow

52° 15′

41° 20′

observer

sun is nearly overhead. The only possibility of seeing a rainbow at that time is from an airplane, a tall skyscraper observation tower, or from the very peak of a mountain. The airplane provides the maximum opportunity for seeing a full 360° rainbow, a sight only few have observed. Artificial bows are produced in the fine spray from a garden hose or in the spray of a waterfall.

Secondary bows are fainter than primary bows because less light energy is present in the doubly reflected ray bundle. This is caused, at least in part, by the aerodynamic shape of the falling drops, whose undersides depart most from a spherical shape when the drops are larger — the largest raindrops look like hamburger buns (certainly *not* like teardrops) with concave undersides. This departure from spherical shape may so distort the path of the lower bundle of rays that there will be no secondary bow, a commonly observed situation. The absence of the secondary bow is thus a convenient test for drop size, indicating the presence of very large drops.

The simple refraction theory just described uses only the rays experiencing minimum deviation from the original path. Any rain shower contains drops of many sizes. Slight color changes due to diffraction of the sunlight may be caused by smaller droplets and sometimes produce additional faint bows inside the principal one.

Light from the rainbow is polarized. Light vibrating horizontally at the top of the bow is much more intense than the light vibrating perpendicularly to it across the bow and it may be as much as 20 times as strong. This can be observed by rotating polarizing sunglasses.

Fig. 30 Double rainbow.

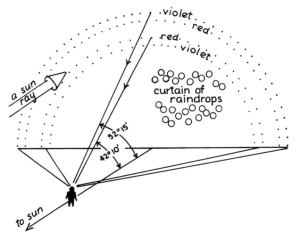

Interpretation of Color Plates

Although each of the 32 color plates has a brief descriptive caption accompanying it, some additional information may be helpful. The following information and comments relate to each of the photographs.

Plate 1. At sunrise and shortly thereafter, this type of cloud is likely to form in the middle levels of the troposphere whenever the air is close to the dew point and unstable. It is a special form of heap cloud called altocumulus castellatus. This cloud forms when the heat of the sun warms the moisture-laden air and starts an upward convective motion. Since there is no continuing source of moisture below the clouds, they do not have a flat base but instead exhibit an overall ragged appearance in their outline and soon disappear.

The formation of these altocumulus clouds with "turrets" in the early morning is a valuable tool in the forecasting of thunderstorm development later in the day. If the air near the ground is moist and potentially unstable, there is an excellent chance that as the day advances and the ground becomes heated, upward convective motions will develop until cumulus clouds begin to form. As they grow larger and their tops reach levels between 5–7 km (16,000–23,000 ft.), they encounter the moist unstable air where the altocumulus castellatus formed at sunrise. Entering this moist unstable layer, they rapidly expand into it, continuing their upward motion until reaching the precipitation (and often thunder-and-lightning) stage.

The turreted altocumulus clouds can be spectacular in appearance but are not often seen except by those who get up at sunrise.

Plate 2. This massive cumulonimbus probably extends well into the stratosphere. The stable layer at the top of the troposphere causes the updraft currents to turn laterally. This carries the ice crystal top of the cloud outward to form a very pronounced anvil shape. The plate shows the striated forms associated with an ice crystal cloud. Also in evidence are the central convective cells that form the active core of the cloud system.

The sun is only a few degrees below the horizon. The cloud is so high that its top half still receives the sun's rays and scatters white light to the observer; lower layers of cloud, in the shadow, are dark. Clouds of this kind, which form over the high plains of Colorado and other states east of the Rocky Mountains, often produce large amounts of hail as they move eastward away from the mountains.

Plate 3. This photograph shows a number of interesting features. The delicate rose-pink color is caused by illumination of the lower surfaces of convective clouds which are in a dissipating stage. The reddish color comes from the light of the setting sun, where the blue end of the spectrum has been scattered so that only the visible red remains to produce the illumination. In the distance a complex wave cloud can be seen, indicating the presence of high velocity winds passing over mountainous terrain which causes the air to move upward, thus producing a cloud at the crest of the wave.

Plate 4. A very large cumulonimbus cloud, whose vertical column is visible on the right-hand side of the photograph, has produced a spreading anvil of ice crystals that has moved to the left. The base of this anvil indicates the presence of instability at middle levels where some altocumulus are forming. The air below has sufficient moisture to form several ragged and broken-up cumulus that are probably being drawn toward the active column on the right. These are in shadow, which indicates that the sun is still fairly high.

In the distance another very active cumulus system is growing. This is in full sunshine and will also probably form an anvil top as the afternoon progresses. These clouds are likely to produce lightning and may develop hail as part of the precipitation process.

Plate 5. Unusually symmetrical lenticular altocumulus gives the appearance of a flying saucer. This degree of symmetry is exceedingly rare although wave clouds are relatively common in the mountains when wind velocity is high. This cloud was photographed in the vicinity of Mt. St. Helens, a volcanic peak in the Cascade Range of the state of Washington. It was forming on the crest of a lee wave that developed several miles northeast (downwind) of the mountaintop. The cloud probably formed when the airflow over the summit of the mountain initiated a lee wave in stable air that overlay a region of instability in which the lower altocumulus had developed.

The remarkable symmetry of this cloud, its resemblance to a flying saucer, and the fact that such clouds may form and disappear in less than a minute, often gives rise to fanciful tales of mysterious objects that appear in the sky.

Plate 6. The undersurface of a layer of altocumulus clouds is illuminated by the setting sun. These clouds were imbedded in a rapidly moving jet stream that had been passing over the Grand Tetons in western Wyoming for more than a day. The moist, unstable air thrust upward by the mountainous terrain produced an ever-changing variety of cloud forms at middle levels in the troposphere. Under more stable conditions, wave clouds would form. In this case, however, the instability prevented stable waves from developing, and as a result, the clouds exhibit irregularities.

Plate 7. A series of cirrocumulus roll clouds are shown forming downwind of the Sierra Nevada. The clouds contain transverse ripples, as seen in the cloud near the top of the photograph. This cloud is similar in structure to those seen in the distance, although they may be somewhat thicker. This type of cloud forms in the lee of the mountains and is indicative of strong winds moving up the western slopes of the Sierra.

At lower levels, ragged cumulus clouds indicate highly turbulent air in the region below the mountain summits. These may be a type of "rotor" cloud that often develops in the lee of high mountains when the air is unstable and some of the fast-moving air in that region forms clouds that look like fractocumulus.

Plate 8. This is the undersurface of a shelf or anvil cloud. This cloud is associated with a large cumulonimbus mass, which is off to the right of the picture. Such a cloud is seen from a greater distance in C.Pl.2. The shelf, or anvil, forms when rising air currents in the body of the cumulonimbus reach a stable layer in the upper atmosphere and are deflected laterally. This cloud did not develop the extreme bulbous or mammatiform structure seen in Pl. 171, which frequently is associated with intense thunderheads that may generate hail or tornadoes.

Plate 9. The rising sun's image seen through a telescope is greatly modified due to the refraction of some of its rays as they pass through air of variable density. This misshapen image occurs immediately after sunrise when the rays from the lower limb of the sun pass through a long section of the dense and often polluted air in the lower part of the atmosphere. In this photograph even the upper limb of the sun is somewhat modified.

Quite rarely at sunrise or sunset (if the view of the sun is not obscured by clouds or polluted air) the first or last glint of the sun will appear as a brilliant green ray — the "green flash." This is also a refraction phenomenon, but it only occurs when the atmosphere is extremely clean and there is little in it to cause diffraction or scattering. The green flash is never seen under the conditions that produced the sun's image in this photograph.

Plate 10. The afterglow seen in the sky after sunset, as shown in this photograph, is produced by the presence of a large number of visible particles in the stratosphere. In this instance, the effect was produced by volcanic ash injected into the upper atmosphere by the explosive eruption of a large volcano in Central America.

When many visible particles are in the air, many of the sun's rays are scattered in a forward direction; this is typical when the particles are larger than several microns. When particles of this kind are in the air of the stratosphere, which is ordinarily extremely clean, they tend to produce very beautiful optical effects. Vivid sunsets occur, followed by an afterglow that may persist for half an hour or more. Dust clouds may remain in the stratosphere for a year or more since the particles have long residence times in that stable region.

Whenever this phenomenon occurs, a bright aureole of scattered light surrounds the sun. The aureole can be seen by holding up an object to block out the sun and observing the scattered light adjacent to it. In very clean air the sky will appear to be very blue, whereas when visible particles are present, it is paler and the diameter of the aureole may be 2–5 times the sun's diameter; this is quite striking during dust storms, volcanic eruptions, or other occurrences that put large particles in the air. Fine particle pollution, on the other hand, does not produce an aureole but tends to diminish the amount of light received from the sun; with high levels of fine (submicron size) particles, it is even possible to look directly at the sun without using a filter. Sometimes under such conditions the sun will appear pink, orange, green, or blue depending on particle size and uniformity.

Plate 11. The sun pillar, 22° halo, and parhelia visible in this photograph are produced by hexagonal ice crystals in the form of plates and columns. The pillar is produced by thin hexagonal *plates* floating in the air with their largest dimension horizontal. The rays of the sun are reflected from the top surface of one to the bottom surface of another, reinforcing the rays in a vertical direction. The halo, on the other hand, is caused by hexagonal *columns;* the diffraction effects are shown in Chapter 4. The halo also is responsible for the enhanced optical effects at the horizontal plane of the sun — the parhelia. These are not very noticeable in this photograph but can be seen by close observation.

Plate 12. This photograph shows a sun pillar without a halo or parhelia — strong evidence that only hexagonal plates are present. It is interesting to note that this pillar is inclined *toward* the observer since it is visible in front of the altocumulus cloud that appears in the center of the photograph.

Plate 13. A fine example of a 22° halo; this one was observed from a ship in the North Atlantic. The halo developed in a vast cloud shield of cirrostratus that was spreading out from the center of a large hurricane several hundred miles to the north. The photograph was obtained by blocking out the sun with a meter stick. Unless the brightness of the direct sun is reduced or eliminated, it is difficult to photograph halo phenomena. Notice also in this photo that the aureole (mentioned in the explanation of Color Plate 10) has a breadth of about 2 suns' diameters.

Plate 14. A halo complex including the 22° and 46° halo, a sun pillar, parhelia, portion of a parhelic circle, and tangential arcs observed at Yellowstone National Park. The ice crystals responsible for these effects were very short hexagonal columns of clear ice that had grown on silver iodide nuclei. These crystals, which were in concentrations of 1–10 per cubic centimeter, had grown to a diameter of 25–50 microns and a similar length, and were confined to a depth of less than 300 m (1000 ft.) above the ground. Typically the low morning sun enhanced the brilliance of the optical effects observed.

Plate 15. A brilliant parhelion, a sun pillar, and a portion of a parhelic circle observed at Yellowstone Park under conditions similar to those described for Color Plate 14. The waters of the Firehole R. reflect some of the effect.

Plate 16. A fine example of a highly localized sunstreak and pillar, with which the faint portion of a parhelic circle combines to form a cross. In addition, a strong 22° parhelion is visible on the right of the photograph. The ice crystals that caused these effects were mostly thin hexagonal plates produced by the dry-ice seeding of the supercooled fogs created by the geysers and hot springs at Yellowstone Park in the vicinity of Old Faithful. The vertical thickness of the ice crystal fog producing these optical effects was less than 300 m (1000 ft.).

Plate 17. The presunrise morning sky with a fan of crepuscular rays is beautifully illustrated in this photograph. There seems to be a layer of thin stratus with cumulus activity in evidence along the horizon. The unevenness of the cumulus profile alternately blocks and allows passage of the sunlight, producing the crepuscular rays that result from light scattering by small particles. The fan shape is illusory; crepuscular rays are actually parallel, only appearing to converge. The yellows of the low sky and cloud blend gradually into the delicate purple of the not yet illuminated sky.

Plate 18. Crepuscular rays from the sun illuminating a late morning fog in the forest. In an effect similar to the one on Color Plate 17, the fanlike array of sunbeams are parallel to each other even though they appear to converge. In this instance the eastern Adirondacks had been immersed since early morning in an extensive layer of fog. As the morning progressed, the fog above the forest was dispersed by the sun's heat and convection, while the fog under the canopy of the forest persisted an hour or so longer. The sharpness of the shadows indicates the highly localized nature of the fog that remained.

Plate 19. A dense mass of mature cumulonimbus and towering cumulus absorb and reflect sunlight. At one cleft between towers, a shaft of sunlight emerges. Small particles of dust act as strong light scatterers. The pronounced crepuscular ray seen against the darker blue of the sky appears brilliantly white. Close inspection shows a shadow cast by the highest tower that is darker than the background sky.

The edges of the cloud mass have a "silver lining" that is again a forward light scattering effect produced by the cloud droplets in the outer, less dense portions of the cloud.

Plate 20. The very large cumulonimbus or swelling cumulus mass of cloud in the lower center acts as an opaque obstacle to the sun's rays and results in the casting of a shadow on the underside of a broken layer of high altostratus and altocumulus. The highest and most distant tower produces the longest shadow, which appears to extend nearly to the cloud itself, suggesting that the top of the tower reaches nearly to the base of the layer cloud. The triangular shape of the shadow is due to the common apparent convergence of parallel lines in the distance.

Wisps of white crossing the shadow are lower clouds still illuminated by the blue sky. The yellows and orange-red tones result from the blue wavelengths of the spectrum being scattered out by passage of sunlight through atmospheric dust.

Plate 21. Virga, or precipitation that evaporates before it reaches the ground, is seen falling from cumuliform clouds hanging over the San Francisco Peaks of northern Arizona. The peaks rise over 4 km (12,000 ft.), and clouds that form over them usually start out as supercooled water droplets. These particles become transformed into ice crystals, and the streaks of precipitation are primarily ice. The low sun to the left catches the virga and bathes it in the yellowish light characteristic of that time of day when the sun is just on the western horizon and when the air is clean of dust. The bases of the clouds are dark, not yet illumined from below.

Plate 22. Light of the setting sun illuminates a mass of virga falling from a dissipating thunderstorm. Quite frequently in the southwestern U.S. the cloud bases are so high and the lower air so dry that the precipitation falling from the clouds evaporates before it reaches the ground.

Plate 23. Sheets and shafts of snow falling from a large convective storm over the Navajo Indian Reservation produce this dramatic scene illuminated by the late afternoon sun. This precipitation is virga, evaporating before it reaches the ground, but it transports cold air downward. The air spreads out below the storm as gusty winds that produce localized dust storms. These effects are common in desert regions, where the exceedingly dry air absorbs the moisture from the falling precipitation.

Plate 24. A condensation trail from a missile traveling the Pacific missile range, known regionally as the "twilight phenomenon" because it is best seen after the sun is below the horizon. Water vapor from the exhaust of the rocket engines freezes on ice nuclei and forms large numbers of ice crystals. In the high troposphere the contrail may grow and develop iridescent colors as shown in the lower portion of this picture. The erratic path of the upper part of the contrail is caused by shearing wind patterns.

Plate 25. When the size of water droplets in cloudy air is less than about 100 microns (0.025 in.), and fog-filled air is illuminated by the sun, a colorless "fog bow" appears. This has the same angle to the sun — 46° — as a colored rainbow, but shows little or no color such as occurs when larger drops are present. Quite frequently, however, one or more colored rings of a glory (see discussion of Color Plate 26) will be seen close to the shadow of the observer. The fog bow can only be produced by liquid water droplets. If the foggy air is made up of ice crystals, no optical effects will be seen opposite the sun except the shadow of the observer. Looking toward the sun through an ice fog is likely to show halo figures, whereas with a water droplet fog a corona will often be seen.

Plate 26. A fog bow on the flat surface of a cloud deck as seen from an airplane. Although much more difficult to see and to photograph than the rainbow or the fog bow at ground level, this opti-

C.Pl.1 An assembly of altocumulus castellatus as they appeared just before sunrise. The presence of such clouds in the early morning indicates that there is moist unstable air at the middle levels.

C.Pl.2 A huge cumulonimbus cloud mass whose upper portions are still illuminated. Note how the rear half of the cloud is in its own shadow. The sun is still above the horizon so the cloud is white.

C.Pl.3 Various types of altocumulus illuminated by the setting sun and viewed from above in a jet plane. An irregular-shaped wave cloud is seen in the distance.

C.Pl.4 Very large cumulonimbus top has been forced to spread out, forming a shaft, in this case, of altocumulus. Observer looks northwest, as the sun illumines the underside of the cloud layer.

C.Pl.5 Flying-saucer-like lenticular altocumulus photographed in lee of Mt. St. Helens, Wash., and marked by unusual symmetry. Cloud sits on the crest of a lee wave caused by airflow over the mountain.

C.Pl.6 An irregular mass of altocumulus carried by a jet stream and made thicker by the lift produced by the Grand Teton Mts. These are illuminated by the setting sun.

C.Pl.7 A complex system of clouds in the lee of the Sierra Nevada. Such clouds are produced by the high velocity winds that often occur in that region.

C.Pl.8 The undersurface, similar to C.Pl.2, of a very large cumulonimbus anvil cloud. Small pouches, or mammata, have not developed to the extent shown in Pl.171.

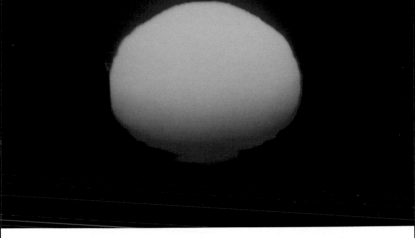

C.Pl.9 Rising sun photographed through a telescope. Distortion of disk of the sun results from refraction. The atmosphere develops small density differences that create the spreading of the image.

C.Pl.10 The afterglow following sunset produced by the forward scattering of light from the sun by particles in the stratosphere. In this case it was caused by volcanic ash from Central America.

C.Pl.11 The combination of sun pillar, 22° halo, and parhelia. Such optical effects are not rare but are rarely noticed. They are caused by refraction and reflection from ice crystals.

C.Pl.12 A sunstreak produced by the sun's reflection from the flat surfaces of hexagonal plate ice crystals. Such crystals fall with their larger axis horizontal, each crystal acting like a tiny mirror.

C.Pl.13 The full circle of a 22° halo that formed within a mass of columnar ice crystals spreading out at the base of the stratosphere; the crystals emerged from the eye of a hurricane.

C.Pl.14 Portions of tangential arcs and circular halos of 22° and 46° formed when a supercooled fog at Yellowstone was seeded with silver iodide, producing short clear columnar ice crystals.

C.Pl.15 A parhelion of 22° produced by seeding a supercooled fog with dry ice. The ice embryos grow into hexagonal columnar ice crystals that refract the sun's rays, causing the optical effect.

C.Pl.16 White cross is formed by the junction of the sunstreak (below) and pillar (above) with the horizontal section of a parhelic circle. Parhelion is visible to the right of Old Faithful.

C.Pl.17 Crepuscular rays at sunrise seen over the ocean. Careful scrutiny shows the presence of trade-wind cumulus to the east; the uneven profile interrupts the sun's beam. Scattering defines the rays.

C.Pl.18 Rays of the sun filtering through a forest and illuminating the water droplets of a morning fog. Although they seem to fan out, these crepuscular rays are actually parallel with each other.

C.Pl.19 Dense mass of cumulus obscures the low-angle sun except for one cleft between towers where it emerges as bright crepuscular ray. Small particles in air scatter sunlight in forward direction.

C.Pl.20 Dramatic shadow is cast by the towers of a distant cumulonimbus or swelling cumulus on an upper layer of high altocumulus and altostratus. Shadow's apparent convergence is due to perspective.

C.Pl.21 Virga over San Francisco Peaks, Ariz., at sunset. High-level clouds frequently generate precipitation in icelike form that falls from the clouds but evaporates well before it reaches the ground.

C.Pl.22 Virga falling from a dissipating thunderstorm illuminated by the setting sun. Such precipitation may reach the ground but only as a very light sprinkle. At cloud base it was probably snow.

C.Pl.23 Unusual display of virga in the form of snow falling from large cumulus clouds in n. Arizona. The evaporating snow led to the development of a violent dust storm at the edge of the cloud system.

C.Pl.24 Track of a Minuteman Missile launched from Vandenberg A.F.B. as seen from Pasadena, Calif. Such condensation trails show the types of air velocity, shear, and turbulence in the high atmosphere.

C.Pl.25 A fog bow. Similar in angle to the rainbow, the fog bow does not show much color during its existence. These are optical effects that appear close to the observer with a low sun near fog's edge.

C.Pl.26 Cloud bow on top of a cloud. Difficult to see as well as photograph, the cloud bow is a very frequent occurrence. Whenever a glory is visible on a cloud, the elongated cloud bow appears.

C.Pl.27 Rainbow at the end of a thunderstorm. Rain falling from the edge of the storm can be seen as a light veil inside of the primary bow, while a heavier rainfall is visible on either side of secondary.

C.Pl.28 Rainbow showing the arc of a primary bow, with the region inside the arc reflecting sunlight from the falling precipitation. Notice the high contrast between inside and outside.

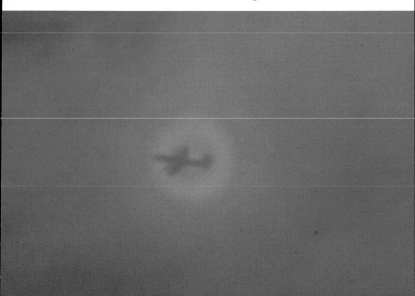

C.Pl.29 A lunar corona produced by the diffraction of moonlight passing through a thin cloud of very uniform liquid droplets. Unlike the glory, it is seen by looking toward the light source.

C.Pl.30 The glory surrounding the plane's shadow is an optical effect produced by spherical cloud droplets. They diffract some of the sunlight back toward the observer as colored rings.

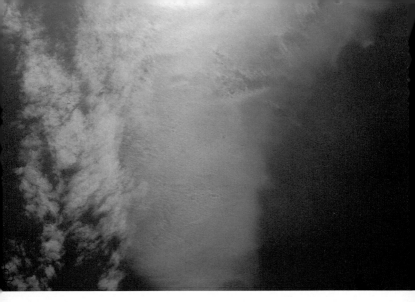

C.Pl.31 The iridescent colors in this wave cloud show it to be composed of very uniform water droplets. It is part of a corona, and the colors are produced by diffraction of the sun's rays.

C.Pl.32 The symmetry of an ice crystal seen with an interference microscope. The thickness and pattern of each ray reflects the uniformity of growth conditions and internal crystalline forces.

cal effect is a common occurrence. The more common optical effect seen from the air is the glory, which is the colored circle surrounding the shadow of the plane. Both cloud bow and glory are proof that the cloud droplets are in liquid form; most often they are also in a supercooled state. If the observer has a Polaroid filter, the optical effects will be considerably intensified.

The fog bow on a cloud deck is most easily seen by watching the cloud surface and noting where the structural detail of the cloud seems to be most clearly illuminated. Because of the geometry of the cloud deck, these fog bows are rarely seen as continuous arches or circles, but rather as a small section of an arch.

Plate 27. The primary and secondary arches of a rainbow are shown against a mountain background close to sunset. Notice the complementary colors of the secondary bow, which shows the reverse in color sequence. The precipitation producing this rainbow is the aftermath of an intensive air-mass type thunderstorm. Cloud bases of these storms are frequently a mile or more above the earth, and as a result, rainbows tend to be seen more frequently and are often more complete in the Southwest and other places where such conditions prevail.

The precipitation that interacts with the light of the setting sun to produce this beautiful effect is seen to consist of shafts of rain with considerable variation in concentration. This is probably responsible for the weakness of the upper part of the secondary arch.

Plate 28. The lower left portion of a circular rainbow as seen from an airplane flying above a dissipating rainstorm. The sun illuminates the tops of altostratus clouds, which remained when the larger cumulus that produced the rain had evaporated. The enhanced reflection from the area inside the bow suggests that drizzle-type drops of uniform size caused the effects photographed. See p. 162 for an explanation of the optical effects of raindrops.

Plate 29. This is a lunar corona—one of the most difficult of atmospheric optical effects to photograph, and an optical phenomenon that is uncommon since it requires a rather unusual set of circumstances. The cloud must consist of uniform-sized liquid water droplets, thin enough to permit the light to reach the observer and yet of enough mass to produce the multiple diffraction and scattering effects that create the corona. This set of requirements is best satisfied by thin wave clouds. Unfortunately such clouds are as a rule of short duration. The amount of light, especially from the moon, requires a time exposure. Quite frequently when such a corona occurs, it has disappeared by the time the camera has been mounted on a tripod and the exposure made. The corona is related to the glory, but the angular displacement of the illuminated disk is generally different. It is a beautiful effect to see.

Plate 30. The colored disk surrounding the shadow of the airplane is called a glory. It is a refraction phenomenon somewhat similar physically to the corona, but is produced by very small liquid cloud droplets close to the spot opposite the sun. The diameter of the ring is directly related to the droplet size. The smaller the glory diameter, the larger the droplets that produce it. Thus the largest glories occur where the clouds are very wispy and thin, and they are therefore not frequently seen by the casual observer on an airplane.

The best glories are seen on the top of wave clouds and stratiform cloud decks, although good ones will occasionally be seen on the top of stratocumulus and even large cumulus. More frequently, however, the larger turbulent clouds have droplets too large and nonuniform in size to produce this effect, or they have been converted to ice particles and thus cannot produce a glory.

When a glory cannot be seen in the antisolar position (where the plane's shadow appears), it is likely that an undersun (pp. 17, 225) will be visible from the other side of the airplane.

Plate 31. Iridescence in clouds is a diffraction phenomenon caused by the interaction of sunlight with cloud droplets of very uniform size. In this example, the cloud is at the crest of a wave in the lee of a high mountain. This wave is quite thin so that the colors extend outward from the sun's location and show at least 2 successive sequences of color, the dominant one being a light blue. The uniformity of droplet size is typical of a "young" cloud, since the droplets are exclusively the result of moisture condensation. With time, the coalescence of droplets will occur, leading to a wider distribution in droplet size. It is likely that the iridescence is part of a corona since it usually appears as a ring of color surrounding the sun. In this instance the cloud is not large enough to provide a symmetrical corona. There is a good chance that a fine glory would be seen surrounding the shadow of an airplane flying over this cloud. While the color in this example is dominated by blue, purplish reds and pale greens are often seen. The color at any given point is determined by the size of the cloud particles and their angular distance from the sun with respect to the observer. Its brightness is enhanced by the number and uniformity of the cloud droplets.

Plate 32. This color plate shows the remarkable symmetry of some ice crystals. The photomicrograph was made using the plastic replica of a crystal in conjunction with an interference microscope. The ice crystal formed on a submicroscopic ice nucleus of pentaerythritol that was put into a cold chamber containing a supercooled cloud. When the ice crystal became large enough, it fell onto a glass slide coated with a cold replication liquid (see Chapter 10). The replica solution covered the crystal with a uniform sheet of plastic when the solvent evaporated. Differences in thickness of the crystal were revealed by this special microscope when the crystal evaporated and its water molecules passed through the thin plastic coating.

Notice how these variations in thickness are symmetrical on each of the 6 rays of the hexagonal crystal, and how this uniformity is maintained down to the tiniest detail. The color effects recorded by the microscope are caused by the same light phenomenon that produces the color in soap bubbles and in oil films on wet streets.

These photographs represent only a tiny bit of the wealth of color to be found in all parts of the atmosphere and its phenomena. While most of the color in the atmosphere is induced by the parallel rays of the sun interacting with air molecules, submicroscopic and visible particles, cloud droplets, ice crystals, snowflakes, hailstones, and raindrops, as single particles or in great numbers, similar but less vivid effects are occasionally produced by the light of the moon as well as by manmade lights. Familiarity with the effects observed by sunlight often explains or draws attention to the less vivid or brilliant effects.

Clouds Seen from Airplanes

One is exposed to a fascinating panorama during an airplane flight, especially in a high-altitude jet aircraft.

In choosing a window seat, it is important to consider where the sun is likely to be during the flight. The traveler has the option of sitting on the side toward or away from the sun except during the middle of the day; each side of the plane has its advantages. Photographs have the best contrast and illumination if taken from the side opposite the sun; this is also the best side for the glory, contrail shadows, the cloud bow, and shock-wave shadow. Some interesting phenomena like halos, parhelia, a green flash, and undersuns can be seen only from the sunny side, however. In selecting a seat, it is of course preferable to avoid those that are over the wing.

To get a good photographic record of the sky, it is important to avoid reflections on the window of objects inside the plane; this is most easily accomplished by holding the camera so it almost touches the window. It should never rest on the window, however, because of aircraft vibration. Although contrast and certain optical effects can be enhanced by using polarizing filters, some aircraft windows are likely to cause unwanted colors in combination with them, so it is best not to use them when flying. (Under certain special conditions even the polarization from the sky at 90° from the sun can produce anomalies.)

For pointers on stereoscopic photography see Chapter 10. The structures present in clouds, particularly cumulus and cirrus, cannot really be appreciated until they have been viewed in 3 dimensions. It also provides a remarkable perspective of the flat top of stratus clouds extending from horizon to horizon, and the hazy, murky pollution dome of the stagnant air of a city. It can be a real psychological shock to cross suddenly from one world where the air is clear and the sky is blue to a completely different world where the air is heavily polluted and there is no blue sky, and realize that this is where many people have to live.

In sharp contrast, the air traveler sometimes has the experience of flying at jet altitudes alongside a gigantic cumulonimbus cloud that extends upward into the stratosphere. Under such circumstances one can ponder the vast amounts of energy released as this type of cloud system penetrates the highly stable region. It also is worth reflecting that the beautiful giant may be releasing a torrential rain shower that will produce a local flashflood, or an on-slaught of hail that may cause agricultural damage; it may even be spawning a dangerous tornado.

The photographs that follow indicate some of the variety of clouds and other objects of interest that can often be seen from an airplane window. The phenomena that are depicted range from relatively stable sea smoke and fog to the rapidly changing features of cumulus towers. At times these latter structures have a breathtaking beauty, especially when illuminated by a low after-noon sun.

If the cloud towers are abundant and extend higher than the plane's altitude — giant storms may extend more than 17 km (11 mi.) above the ground — the observer will note that the pilot may maneuver the plane off of a straight course. He does this to avoid the intense convective parts of the clouds that may cause turbulence, hail, or lightning. The decision to do this is based on information from a small radar in the nose of the plane, which provides him with a visual display of the most intense parts of the storm ahead of him.

Under normal circumstances most clouds are below the altitude flown by jet airplanes, which ranges from 9 to 12 km (31,000–41,000 ft.). At these altitudes it is possible on a clear day to see objects more than 320 km (200 mi.) away if the air is clean.

There is an endless variety in cloud patterns and combinations. All clouds will appear different when viewed from the perspective of an airplane, and some, like wave and cirrus clouds, particu-larly so.

Pl.181 A boiling mass of convective heap clouds that will soon merge to form a large storm system. If they reach flight level, radar will be used to avoid them as such air is very turbulent.

Pl.182 Valley fog outlines the upper watershed of a large river system. The distant view is restricted by a humid haze that forms when hygroscopic particles acquire moisture.

Pl.183 The organized pattern of a frontal system showing cloud bands that will merge into massive towering convective cumulus. This in turn will lead to a convergence of moist air from lower levels.

Pl.184 The appearance of a fast-moving frontal system. The front is moving from left to right with the strongest winds at middle levels. This accounts for the lag of the topmost part of the anvil clouds.

Pl.185 Early formative stage of an air-mass type of thunderstorm. Vigorous growth of towering cumulus proceeding in all areas. Two earlier towers have encountered stable air and formed anvils.

178

Pl.186 The final stage in the development of afternoon thunderstorms. The large cumulus towers, which formed earlier in the day, have flattened out to form extensive anvil tops of ice crystals.

Pl.187 This stratus cloud deck, which has completely covered the valleys of this mountain region, is probably a fog that reached the ground. Such clouds are often supercooled and could be removed by seeding.

Pl.188 A solid deck of stratus clouds covers a mountainous area with only the peaks visible. At the lower left of this view, clouds tumble through a mountain pass while larger waves develop nearby.

Pl.189 Lines of convective clouds that are beginning to show vertical growth. It is likely that later in the day the larger clouds will merge and bands of rain will spread across the region.

180

Pl.190 Gusty winds cause finely divided soil particles to become airborne during dry periods. Such soil may be transported for long distances. This is one of the major sources of ice nuclei.

Pl.191 The smoke from a very small forest fire has reached a stable layer in the atmosphere and is spreading horizontally, probably in a circular pattern. Smoke from another fire is beyond.

181

Pl.192 Ripples produced in a thin layer of stratus by a very localized disturbance. This effect was induced by the convective plume from a single cooking fire in a tiny village shortly after sunrise.

Pl.193 Solid deck of stratocumulus extends for hundreds of miles over Atlantic Ocean. The several shear zones are produced by discontinuities in the air-flow temperature or other factors.

Pl.194 An extensive deck of altocumulus clouds of the type that often form at the base of a jet stream. Such cloud regions may form parallel patterns that become oriented at right angles to the flow.

Pl.195 Small fair-weather cumulus. Such clouds form when moist air rises from the ground as it is heated by the sun. The convective column forms a cloud when the moist air reaches its dew point.

Pl.196 A continuous layer of stratocumulus clouds. Such clouds tend to be extensive; they form when moist, slightly unstable air is forced upward by convergence or other large-scale air movements.

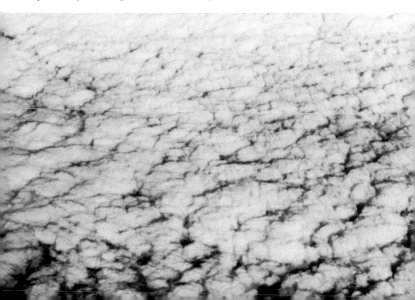

Pl.197 While this looks like a thin cloud, it actually is smog from manmade pollution. Many western mountain valleys are susceptible to this type of problem, especially in the wintertime.

184

Pl.198 The circles on the ground are land irrigated by a slowly moving spray rig that receives water from a well at the center of the circle. Patches of stratocumulus drift across the irrigated area.

Pl.199 A stratiform deck of clouds affected by a strong wind that is producing waves similar to those often seen on water. A band of higher clouds shows that the wind effects are widespread.

Pl.200 Cloud waves that resemble similar structures in the ocean. When vertical motion is restricted but there is a strong wind at the top of the cloud, such billows form and are quite turbulent.

Pl.201 Precipitating trade-wind cumulus clouds in the foreground are in sharp contrast to the extensive areas of layer clouds in the distance. Trade-wind cumulus often have ragged bases.

186

Pl.202 Wave cloud with ripples induced by a land mass that causes the moist maritime air to rise and cool, thus forming the cloud. The effect of the land extends both up- and down-stream.

Pl.203 Mountain-induced wave clouds in formative phase are the dominant feature, while in the distance are stratocumulus of irregular formations. Cirrostratus clouds form the overcast above.

Pl.204 A wave cloud that has formed over a mountain. Such clouds are produced when strong winds force air over a barrier. The cloud may produce ice crystals that stream downwind for long distances.

Pl.205 Cumulus of the northeast trade winds. They often form polygonal patterns, with the air moving downward in the center and rising along the edges to form towering clouds. They frequently rain.

Pl.206 A very complex system of clouds. Cirrus and cirrocumulus are in the high sky. Small cumulus are near the ground, while towering cumulus and stratiform clouds occupy the middle levels.

Pl.207 Towering cumulus forming over warmer land at the edge of the colder ocean. The air at sea level over the water is calm, but very likely there is a wind under the cloudy region.

Pl.208 Convective clouds, some of them quite large, form above warm land. The higher clouds have produced ice crystals or mist particles which reflect light to a very different degree.

Pl.209 Classic example of a heap cloud showing all stages of formation except fully developed anvil. A mature tower is just starting to form an anvil at right. A new tower is in the foreground.

Pl.210 Looking down at top of "boiling" mass of convective cumulus. This view clearly shows why such cumulus are often described as cauliflower clouds. Convective towers extremely turbulent.

Pl.211 Classic examples of towering cumulus, one of which has formed an anvil top. In several places thin, darkish shelf clouds have emerged from the cumulus towers to spread in horizontal layers.

191

Pl.212 The beginning of a tropical rainstorm on edge of Pacific. Heavy precipitation has already started. Storm has produced a strong outflow of cold air, which produces more cloud.

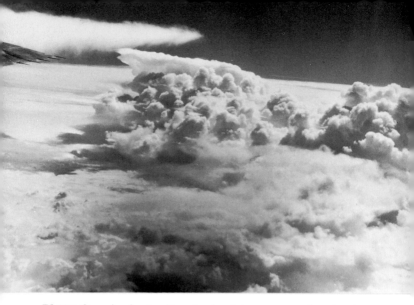

Pl.213 Oceanic clouds often develop into very complex structures. Cumulus in foreground are forming at several levels while larger cumulus project above a solid deck of stratus beyond.

Pl.214 A classic example of large cumulus that have become completely glaciated. Even the smaller clouds visible in the corridor appear to have been converted to ice crystals.

Pl.215 The plane is flying below a huge glaciated anvil. Similar clouds can be seen nearby and farther off, surrounded with towering cumulus that later will also form anvils.

Pl.216 Edge of a large frontal system consisting of towering cumulus, all of which have formed anvil tops. Flow of air into storm is marked by low cumulus at edge of larger clouds.

6

Severe Storms

Severe storms are impressive and terrifying natural phenomena. They are unusual manifestations of the organized energy from the sun received by the earth, oceans, and atmosphere.

They range in size from the large cyclonic storm that affects many thousands of square kilometers to the savage tornado that devastates part of a town. Most severe storms are only moderately intense, but from time to time (from a few times a year to a few times a century) the supply of moisture, the convection patterns, the convergence flow patterns, and the instability of the atmosphere are just right to allow a "giant" storm, in size or intensity or both, to develop. When this occurs the storm and its effects overwhelm the normal precautions taken for the protection of life and property, and a disaster occurs. A few such cataclysmic storms are likely to be the cause of many of the scars of land erosion or widespread timber blowdown to be seen in various parts of the world.

Considerable progress has been made in recent years toward improving the forecasting of severe storms and providing storm alerts, watches, and warnings by radio and television. With reliable forecasts and adequate warning, protective steps and safety measures have reduced to a remarkable degree the human death toll from severe storms. Nature, however, is very capricious; there are times when even the most elaborate of plans and warnings are inadequate or become inoperative. But the weatherwise individual can often provide valuable information to his or her family and neighbors by understanding weather signs and taking proper precautions.

The cumulonimbus cloud is a veritable weather factory. Out of individual cumulonimbus clouds can come at least 5 distinct severe atmospheric effects: (1) thunderstorm with associated heavy winds, (2) cloudburst, (3) hail, (4) tornado, and (5) waterspout. Cumulonimbi that produce severe weather are awesome giants of clouds, extending sometimes to the top of the troposphere — roughly 12 km (40,000 ft.) — and occasionally into the lower stratosphere. Clouds of this size do not occur unless there is an adequate supply of energy, and this means the air mass must be loaded with copious amounts of water vapor distributed through a deep layer. Furthermore, the right set of conditions must exist to trigger the cloud's development, and there must be no inhibiting factors that prevent its growth to full maturity, such as inversion layers in the middle levels of the atmosphere.

The explanations that follow can provide only a general account of what is going on during severe storms. The reader should refer to the numerous texts in meteorology for more details.

Thunderstorm

As it forms, a cumulonimbus thunderstorm cloud develops a separation of electrical charge, with a center of positive charge in the frozen upper portion of the cloud and center of negative charge near its base. Precisely how this happens is still not known with certainty, mainly because of the extreme complexity of the problem and the difficulties of making enough measurements on the same storm so that reasonable conclusions can be drawn. Although it has been done, it is very difficult and dangerous to make measurements inside the cloud because of inaccessibility, extreme turbulence, and the strong electrical fields themselves. It is known from laboratory work that breakup of raindrops, freezing, ice splintering, frosting, and rubbing effects — crystal on crystal and drop on drop — are some of the events producing charge separation.

The earth is a negatively charged body, but under a thunderstorm the negative charge is pushed away by electrostatic repulsion of the underside of the cloud (charged negatively), and a positive charge is induced that becomes stronger as the cloud's negative charge center strengthens. Positively charged ions move up to the top of elevated objects and attempt to establish a flow of current, but are prevented from doing so by the air, which is a poor conductor of electricity.

Lightning occurs when the electrical field becomes so strong — as high as 1 million volts per meter — that the insulating capability of the air breaks down. The lightning stroke is a sudden flow, primarily of negative charge, from the cloud to the ground through a channel initiated by a complex series of "leader strokes" followed in rapid succession by a ground-to-cloud "streamer" carrying a positive charge. These continue until the strength of the electric field is reduced — in perhaps 1 second.

The path of the lightning stroke carries peak current reaching 200,000 amperes or more, and the associated temperatures may rise to millions of degrees. As a consequence of this heating, there is explosive expansion and contraction of the air, producing the pressure pulses that are heard as the crash and rumble of thunder.

Because of the speed of sound, the distance in miles to a lightning stroke can be quickly estimated by counting the number of seconds between the sight of the lightning flash and the sound of the thunder, and dividing by 5 (if flash and sound are simultaneous, it was close).

Lightning comes in many forms. *Streak lightning* is the kind most frequently seen — a single or multiple line from cloud to ground; *forked lightning* shows multiple channels; *ball lightning* (a controversial form) appears as a luminous globe that maneuvers like a flying saucer; *heat lightning,* seen along horizons in hot weather, is the reflection of lightning occurring beyond the horizon. St. Elmo's Fire is a phenomenon that occurs when the electrical potential produces coronae from grounded objects, most typically metal ones.

There are about 1800 thunderstorms in progress over the earth's surface at any moment, and strokes hit the earth 100 times each second. The thunderstorm performs the vital function of returning to earth much, if not all, of the negative charge it continually loses by leakage to the atmosphere.

According to data assembled by the National Center for Health Statistics, the annual death toll from lightning is greater than for tornadoes or hurricanes. About 150 Americans are killed by lightning per year, and 250 treated for injuries. Property damage is high.

Note: Persons struck by lightning receive a severe electrical shock and may be burned, but they *carry* no electrical charge and can be handled safely. A person "killed" by lightning can often be revived by prompt mouth-to-mouth resuscitation, cardiac massage, and prolonged artificial respiration. In a group struck by lightning, those apparently dead should be treated first; those who show vital signs will probably recover spontaneously, although burns and other injuries may require treatment. Recovery from lightning strikes is usually complete except for possible impairment or loss of sight or hearing.

Cloudburst

This is a sudden burst of excessively heavy rain from an extremely vigorous cumulonimbus cloud. Such storms are characteristic of summer conditions. When they occur over high country, such as the Rocky Mountains or the neighboring terrain, the rain may be suddenly channeled down an otherwise dry streambed and produce a flashflood. This can present real danger to campers or picnickers, and to small communities whose houses are built too close to the stream channel.

Hikers and campers should be very careful about choosing an overnight camping site, particularly if any cumulonimbus activity is noted.

Every community should have some sort of warning system for flashfloods (5 minutes notice may save many lives) especially if it is located downstream from a dam that might become endangered by sudden and excessive upstream rainfall.

Hail

Every so often a severe cumulonimbus generates hail. There is evidence that hail forms rapidly. An ice pellet gets caught in a powerful updraft, grows as it collides with supercooled water drops, falls, is caught again, and still again, and finally gets into a downdraft where it and other hailstones fall from cloud to ground. There is new evidence that some hailstones grow by "sitting still" in the updraft and gathering the supercooled drops that strike them.

Perhaps the average hailstone in a midwest storm is marble-sized; in the more severe storms the stones are the size of golf balls. Excessively severe storms generate some stones the size of baseballs. Hail can and does cause considerable structural and crop damage; occasionally livestock and humans are also killed.

Substantial efforts are being made to study the hailstorm in hopes of altering growth patterns by weather modification techniques to prevent large hail from falling from the clouds, but much more work needs to be done. The major effort is to produce many smaller hailstones instead of a few large ones. This is accomplished by seeding the storm in its early stages with many more ice nuclei than are present under natural conditions. This action may also reduce the size of the storm.

Tornado

The cause of tornadoes is known generally but not precisely. The ingredients for a tornado-spawning cumulonimbus are: a southerly flow of warm, very moist air, topped by a westerly flow of cool, dry air, and a squall line whose convergence sets off convective clouds. Tornadoes usually, but not always, move along in a southwest-to-northeast direction. The funnel dips down from cloud to ground, becoming visible as moist air moves into the region of sharply lowered pressure and condenses, and as the vortex sucks in debris from the ground.

Tornadoes are difficult to forecast with precision. However, most communities in tornado-prone regions are covered by the Tornado Watch of the National Weather Service and receive advisories by radio and television.

People caught in the open should try to find refuge in a ditch or culvert, and keep down to avoid injury from flying debris. Those interested in scientific observation might try to make a written and photographic record of the behavior of the funnel.

In the home, a basement wall, or preferably storm shelter, will provide the best protection; crouching under a heavy table will help some; do not go upstairs.

Windows must be opened to prevent the house from exploding as a result of the suddenly lowered outside barometric pressure near the funnel. Gas and electricity mains should be turned off, and flashlights and a transistor radio readily available. Refer to Appendix 19.

Waterspout

The waterspout is a first cousin of the tornado, forming from an active cumulonimbus cloud over the water, and developing a long trunk-shaped funnel that dips down to the water. Because of associated strong winds and resulting disturbance of the water surface, waterspouts should be avoided if at all possible by ships and particularly by small boats.

Tropical Storm

Tropical storms (known by the local names of hurricanes, typhoons, cyclones) breed in the low-latitude belts over the oceans, primarily from 5° to 15° north or south latitude. Westward-moving regions of unstable air produce areas of intense vertical growth. Colonies of cumulonimbus clouds generate heavy rain squalls. In a way not fully understood these colonies start to become organized as a circulating wind system, counterclockwise in the northern hemisphere and clockwise in the southern hemisphere. As growth and organization proceed, the storm develops a ring of intense convective clouds surrounding a central, relatively cloudless region at the storm center known as the eye of the storm (the relative absence of clouds at the eye is accounted for by the sinking of the air). Surrounding the central core are radiating bands of large cumulus that take on a pinwheel effect when viewed from high above.

Northern hemisphere storms move out of the central tropical Atlantic and Pacific toward the west, generally guided by the upper wind stream. As they reach the western end of the ocean basin, they tend to recurve to the north and then to the northeast. However, the guidance is a critically balanced matter and some storms' paths are erratic in the extreme, making northward starts and retreats several times, most of the time moving on to the west.

Severe tropical storms can generate surface winds up to 160 km/hr. (100 mi./hr.) and are therefore dangerous if only because of the winds. However, the strong winds also produce tremendous seas in advance. If the impact of a hurricane on the land takes place at a time of extreme high tides, then there can be massive flooding. Furthermore, the circulation of the storm pumps great amounts of warm, moist air ahead of it, and the result of the condensation of this water vapor and consequent precipitation can produce flooding as well.

A hurricane, or typhoon, involves a massive release of atmospheric potential energy. At this stage in history, little can be done to modify such a storm system, though this is a subject of current study, and beginnings have been made.

The generation, development, and movement of these storms are today kept under continuous surveillance by weather satellites and by weather radar stations that are particularly concentrated at the southeast perimeter of the U.S.

An effective warning system has made it possible to order wholesale evacuation well in advance from land regions the storm is to strike. Though such storms still produce extensive property damage, loss of life is now reduced to low levels.

A note of precaution: Follow the instructions of the local Severe Storms Warning Service or other competent authority.

Blizzard

Following a winter frontal storm, which may have deposited from 2–6 in. or more of snow, the airflow swings from the south or southwest to the west or northwest and often the wind produces blizzard conditions. A blizzard consists of strong winds at ground level sometimes exceeding 80 km/hr. (50 mi./hr.), containing airborne snow and ice particles in blinding proportions. The winds are gusty and persistent, producing massive snow drifts and cornices in the lee of mountaintops, road cuts, houses, and other obstacles. Under some conditions a snow structure known as a "sastrugi" forms. It resembles a sand dune but is a compacted, rippled, icy substance, the bane of polar travelers. Although a sastrugi resembles a permanent structure, it, like a sand dune, is in the constant process of transformation as long as the wind blows.

Whiteout

One aspect of a blizzard is the "whiteout" which, because of the high density of airborne particles, restricts visibility to a few meters or less. A whiteout can also be produced by much smaller particles such as a liquid-droplet or an ice-crystal fog. Under whiteout conditions the sense of gravity can nearly disappear, as mountain skiers learn to their dismay.

A special form of whiteout is the ground blizzard, in which the restricted visibility has a vertical thickness often of less than 1.8 m (6 ft.). Polar explorers protected by tent or igloo quickly learn to occasionally check the depth of the whiteout swirling around their temporary habitation. Ground blizzards sometimes persist for many days.

Ice Storm (sometimes called silver thaw)

In the middle latitudes of the continents, the ice storm often develops in winter when warm moist air encounters a cold air mass and slides up over the colder, denser air. When this happens precipitation often forms, generally by coalescence, although the raindrops may be mixed with ice crystals falling into the warmer air from cirrus levels. The rattle of sleet on the windowpanes is the harbinger of the ice storm. Generally the sleet (frozen raindrops) is mixed with liquid drops. When the wet precipitation falls into the cold air at ground level, it freezes upon contact with trees, wires, shrubbery, and any other objects colder than 0°C (32°F). The resulting sheath of ice is beautiful but often devastating.

After the ice thickness reaches 1 cm (⅜ in.) or so, the unnatural burden begins to snap power lines, branches, and similar objects. Roads become like skating rinks and are extremely hazardous.

Fortunately in most instances the warmer air slowly displaces the cold air, the ground temperature rises, and the ice melts. Occasionally this does not happen — after several days the precipitation may end but the air remains cold. If a wind develops, as it often does under these conditions, great damage ensues as the overburdened ice-coated objects break, smashing other things as they fall. Of particular import is precipitation in the form of a misty rain; when this occurs practically all of it freezes upon contact without falling to the ground. Thus the damage is frequently extensive.

Lake Effect Storm

This is a special type of snowstorm that occurs in the lee of large bodies of water. While most frequent downwind of the Great Lakes in the northeastern United States, such storms are common in northern Japan downwind of the Japan Sea and in any other place where cold air has a trajectory of 16 km (10 mi.) or more over open waters that are considerably warmer than the air.

Following the passage of an intensive low-pressure storm across such a region, the cold air streaming from the Polar regions and crossing the warm waters of the lakes or sea picks up moisture and forms extensive cloud streets consisting of large towering cumulus. These often resemble a squall line but the blizzardlike conditions beneath the cloud street may persist for several days. It is not uncommon for such a storm to deposit 1.2–1.5 m (4–5 ft.) of snow in bands roughly 20 km (12 mi.) wide extending for 75 km (47 mi.) or more in the direction of the prevailing wind. Fierce turbulence, heavy drifting, and near-zero visibility occur during these storms. In areas that are the hardest hit, the winter snow depths may exceed 6 m (20 ft.). In parts of northern Japan, villages are completely covered with the snow. This necessitates tunnels leading from the houses to the narrow channels in the streets, where traffic moves with difficulty.

During lake effect storms the narrow precipitation bands often sweep across the region, producing a remarkably uniform deposit over considerable areas of the countryside. At other times the band persists and may extend for several hundred kilometers downwind of the lake.

Flooding

Flooding from excessive precipitation or the melting of accumulated ice and snow is an annual phenomenon on most of the major river systems. The annual deposition of silt on the river bottoms was a factor of great agricultural importance in the past, prior to the use of chemical fertilizers. Local flooding in the form of flashfloods or "gully washers" results from intensive localized convective storms (as noted earlier). If the heavy precipitation from such a storm persists over a small watershed for more than an hour, this type of flooding is likely to occur. At other times a massive frontal storm that becomes stalled, or the intensive rains from a hurricane that moves inland to deposit 10–25 cm (4–10 in.) of rain within a day or two, will produce extensive flooding with its attendant property damage and loss of life.

At other times a moderate rainstorm that persists several times longer than the average can also create extensive damage, especially in large regions of flat land or in mountainous areas where smooth rock is overlaid with thin soil. When the soil in these places becomes saturated, it may suddenly slip from its precarious position to produce a devastating landslide consisting of a mixture of soil, rock, trees, and humus that may form a temporary and highly unstable barrier in the streambed. The rising water then builds up enough pressure to break the dam, the rush of water and debris causing cataclysmic changes to the terrain with possible loss of life and extensive damage to property.

Dust Storm

Like a blizzard of snow, a dust storm is driven by gusty winds that occur when strong pressure gradients develop across a dry barren region. Dust storms may be so intense that they generate enough electrical potential to produce lightning. The *haboob,* a dust storm of central Africa, and the storms that occurred in the mid 1930s to produce the Dust Bowl of the western plains in America are examples of this type of atmospheric phenomenon.

The dust storms of the Sahelian Sahara, an area extending eastward from Senegal, can be seen by satellite to extend across the Atlantic Ocean to the Caribbean Sea or into central Europe, depending on the direction of the prevailing winds.

The soil fragments that are airborne tend to be rounded grains of quartz and other hard rock having a cross section of 5–25 microns. They restrict visibility and may extend upward to the base of the stratosphere.

Katabatic Winds

Under special atmospheric conditions, the air flowing over a mountainous area moves downward following the contours of the lee side of the mountain slope, warming but drying as it descends. This air sometimes behaves like an avalanche, gathering speed as it descends. Reaching the flat lands below, it streams across them, sometimes reaching velocities of 45 m/sec. (100 mi./hr.) in localized, exposed areas.

In its milder forms, this type of descending air and the resulting warm, drying, persistent wind has many localized names. It is the chinook of the western prairies, the foehn of alpine Europe, the mistral of southern France, the Washoe zephyr in the lee of the Sierras, and the Santa Ana of southern California. Because such winds are warm and dry and often occur when snow is on the ground, they have an insatiable appetite for moisture and cause it to disappear as though by magic. These hot, dry winds are known also to produce physiological effects in human beings, causing nervousness, instability, and a general sense of malaise.

Firestorm

One of the most awesome of natural atmospheric phenomena, similar in many respects to the tornado, is the firestorm. This often develops when a forest wildfire becomes organized by a cyclonic circulation. The highly unstable air (greatly exceeding the dry adiabatic rate) pulls in the surrounding air and develops a massive fire of terrifying proportions. An organized firestorm creates its own localized wind pattern which often becomes so violent as to fell trees, tear burning limbs from them, and scatter embers from the upper levels of the convective column far and wide, thus starting new fires.

A blowup is a wildfire that is suddenly intensified by a strong wind, such as a low-level jet stream. This may become a firestorm if a convective column develops and generates its own inflow winds. Classic examples of blowups can be found in the files of the forest services of the United States, Canada, and Australia. The Peshtigo Fire of northeastern Wisconsin occurred simultaneously with the great Chicago fire. A similar fire was the great blowup of western Montana and eastern Idaho that occurred in 1910. Effects of this vast fire can still be seen in the higher slopes of the Bitterroot Mountains of that region.

Most of these blowup fires seem to occur when an intensive jet stream is over the region. Field evidence indicates that extremely dry subsiding air with a persistent wind velocity of 15–25 m/sec. (33–56 mi./hr.), probably in the form of a secondary jet current, initiates the impulse that produces the blowup fire and the firestorm.

Pl.217 Torrential rain from a large thunderstorm. Such precipitation may exceed a rate of 4–5 cm an hour but generally lasts for only 20–30 min. in any one place. Rain may go on for several hours.

Pl.218 A nighttime lightning storm photographed by the light from the discharges occurring within the cloud system. Such local storms are quite spectacular but are uncommon at night.

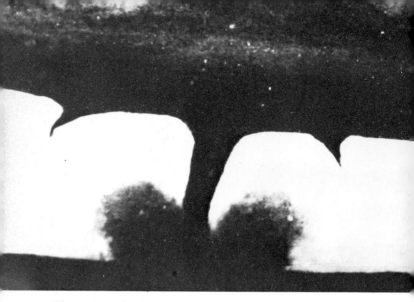

Pl.219 An early photograph of a tornado; occurred in 1884 and was seen for nearly two hours. It has unusual symmetry and this probably accounts for its exceptionally long life.

Pl.220 Typical appearance of a hail shaft. Because of their icy nature hailstones reflect light more than rain. Falling hail can often be identified by the whitish and streaky look.

Pl.221 Mature stage of large air mass thunderstorms. Heavy rain falls in the far distance; nearer, strong winds carry ground level air up into active convective part of storm.

Pl.222 Heavy snows falling from clouds that have formed in the lee of the San Francisco Mountains of Arizona. Streamers of ice crystals move downwind of the mountain.

Pl.223 Leading edge of a cold air mass moving into a zone of warmer air. Precipitation fills air in front of turbulent cloud and will end shortly after windy air arrives.

Pl.224 Highly turbulent clouds along leading edge of a cold front. The cold unstable air pouring over mountains engulfs warmer, more moist air and forces it upward where condensation occurs.

Pl.225 The beginning of precipitation from leading edge of a large cumulonimbus thunderstorm. The lower stratocumulus will be drawn into convective center of advancing storm.

Pl.226 Mammatiform clouds drooping from sloping anvil top of a large cumulonimbus storm whose center is at left in photograph. Clouds in the distance are from an adjacent storm.

Pl.227 A very complex storm system containing every variety of cloud with zones of heavy precipitation. This assembly of clouds is typical of those that make up large cyclonic storm systems.

208

Pl.228 The type of mixed clouds that make up the structure of the spiral bands in a hurricane. Hazy air containing salt particles from the sea restricts visibility. The low clouds feed the cumulus.

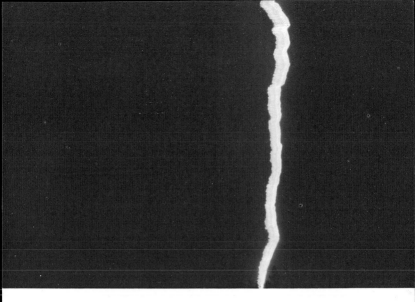

Pl.229 Remarkable view of a lightning stroke occurring about 300 meters from camera. Strong wind has carried the ionized air from right to left, disclosing seven separate strike channels.

209

Pl.230 A very complex multiple lightning strike originating in a single channel but branching into four separate shafts, of which three hit the ground. Storm occurred in western Montana.

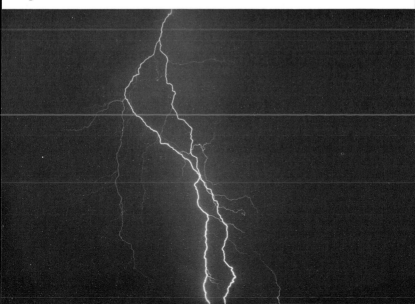

Pl.231 Nighttime radiation cools slopes surrounding this mountain village. Cold dense air flows down into valley, displacing warmer air and forming an inversion that traps polluted air.

Pl.232 Typical pollution pall above large city, mostly formed by coagulation of submicroscopic particles that come from a wide variety of sources ranging from autos to incinerators.

Pl.233 Convective cloud of smoke and gases from a large forest fire started by lightning. The top of the cloud consists of water droplets that form when the rising moisture reaches its dew point.

211

Pl.234 One of the primary sources of atmospheric particles of natural origin. A forest fire ignited by lightning burns along a wide front in Glacier Park of the northern Rocky Mts.

Pl.235 Sharp outline of sun or moon seen through a cloud is evidence that cloud consists entirely of small water droplets. Colored corona occurs if droplet size is uniform.

212

Pl.236 When outline of sun or moon is fuzzy or diffuse, as seen through a piece of ground glass, the air contains large numbers of ice crystals. This often occurs as a warm front approaches.

Pl.237 Stable air restricts vertical growth of a smoke plume that is carried as a thin streamer away from the source. Diffusion and weak convection slowly spread smoke into atmosphere.

213

Pl.238 The complex structure of a carbon smoke particle consisting of thousands of submicroscopic units that have chained together. Static electricity causes this formation.

Pl.239 A mass of airborne particles illuminated by the setting sun. Light scattered by the particles is accentuated by the shadows cast by cumulus towers.

Pl.240 A series of condensation trails all made at about the same level in the atmosphere. Most recent is curved and its pendules of ring vortices show the air is quite stable and homogeneous.

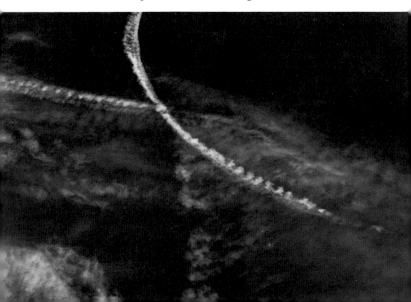

7

Particles in the Atmosphere

There are tremendous numbers of particles suspended in the atmosphere. They vary greatly in size: the smallest are gaseous clusters and ions and submicroscopic liquids and solids; somewhat larger ones produce the beautiful blue haze of distant vistas; those 2 or 3 times larger are highly effective in scattering light; and the largest consist of such things as salt crystals, rock fragments, and the ashy residues from forest fires, volcanoes, or incinerators. There are also nature's living or semiliving particles — spores and pollen grains, spider webs, mites and other tiny insects, plant tissues, diatoms, and even leaves and tumbleweeds. There is also a bewildering variety of particles produced by automobiles, refineries, steel mills, and myriad other human activities, all of which are considered manmade gases and particles. The numbers in which they are concentrated in different parts of the atmosphere vary greatly — from more than 10 million per cubic centimeter to less than 1 per liter (0.001 per cc). Excluding the particles in gases as well as the leaves and tumbleweeds, sizes range from 0.005 to 500 microns, a variation in diameter of 100,000 times. Size range of these particles, combined with their concentration, is a measure of the air quality of a region.

The largest *number* of airborne particles is always in the *invisible range*. These numbers vary from less than 1 per liter to more than a half million per cubic centimeter in heavily polluted air, and to at least 10 times more than that when a gas-to-particle reaction is occurring.

In the very clean parts of the atmosphere there are so few particles that they are difficult to see, collect, or measure. In these clean areas, there are generally less than 300 particles, including the submicroscopic ones, per cubic centimeter, and the ratio of visible to invisible particles is about 1 to 3.

At the other end of the scale — the air of a large city that contains a fantastic variety of particles from many sources — the concentration often exceeds 100,000 per cubic centimeter. There may be more than 3000 *visible* particles per cubic centimeter, making the ratio of visible to invisible particles about 1 to 300. Only the largest of these particles have an appreciable settling rate, so that the majority of them float for a long time and move with the air. Most are eventually removed from the atmosphere by snow or rain.

Despite more than a century of research on the fine particles in the air, there are still many unanswered questions concerning the relationships between the trace gases in the atmosphere and the submicroscopic particles they produce through intermolecular reactions, the larger particles that form by agglomeration of finer ones, the breakup of the largest ones, and the production of bacteria, spores, pollen grains, fibers, and other cellular structures. Fig. 1 in Chapter 1 (p. 4) shows the relative sizes of these different kinds of airborne particles.

The Mechanisms That Produce Airborne Particles

Atmospheric particles are formed in 3 basic ways. The first and most important, the condensation of heated gases, produces the lārgest *number* of particles. These gases may come from many sources — automobiles, steel plants, incinerators, forest fires, or volcanoes. The smallest particles of this type are ions and molecular clusters ranging from 0.001 to 0.005 micron in diameter. Particles of 0.005 micron are assumed to have reached a stable size. Somewhat larger particles, i.e., those formed through Brownian coagulation in highly concentrated vapors, range up to the faintly visible (between 0.1 and 0.3 micron) and produce the beautiful opalescent blue haze of the deserts, mountains, and canyons. Still larger ones (0.3–1 micron) are responsible for the smoky gray hazes, some of which may have grown by the condensation of water molecules on such smaller hygroscopic nuclei as salt particles. All of these, except salt crystals, are the result of gas-to-particle reactions. Such haze particles, which affect visibility, rarely exceed a diameter of 1 micron unless they are hygroscopic or are in air that has reached the dew point. When hygroscopic, they begin to acquire water molecules at relative humidities between 50 and 70%. Under dew-point conditions, water molecules condense on them to produce a fog with droplets that may exceed a diameter of 10 microns. When such a droplet evaporates, the particle on which the moisture condensed reverts to its original size, although it may be somewhat larger if other particles contacted it while the droplet floated in the atmosphere.

The second basic mechanism that produces the largest airborne particles is the fragmentation of larger particles by mechanical breakage, abrasion, corrosion, or deterioration. The particles may be as large as 100 microns and consist of fly ash from power plants, catalytic powders from refineries, cement or other mineral dust from industrial grinding or quarrying processes, or the ashes from incinerators, forest fires, or other burning. Such free-floating particles are rarely less than 1 micron in diameter, since any smaller ones formed by these processes are likely to be stuck to larger ones by electrostatic, chemical, or other binding forces.

The third basic source of airborne particles combines the processes of condensation, coalescence, and agglomeration to form air-borne stable droplets, irregular-shaped aggregates, or chainlike clusters held together by electrostatic or mechanical forces.

The Size Range of Airborne Particles

The size of airborne particles is not easy to comprehend because of the wide variation that occurs. Particle sizes range from those that are quite invisible to the unaided eye to those that restrict visibility.

It is convenient to consider all particles spherical for purpose of measurement; in fact, many of the particles are not spherical but consist of irregular shapes, chains, and clusters. These particles have a wide variety of shape, density, and refractive ability.

The actual dimensions of airborne particles range from 0.001 to 100 microns. A 100-micron particle has a cross section slightly larger than a human hair. A particle 1000 times smaller in diameter (0.1 micron) is just visible when illuminated with a stray beam of light although its shape cannot be seen except under an electron microscope. This is the type that produces the blue haze of such places as the Grand Canyon and distant vistas of forested mountains far removed from the polluted air of cities.

Particles 5 or 6 times larger in diameter (0.5–0.6 micron) are of the size that scatters light most effectively. These particles produce a smoky haze. The number of airborne particles larger than a micron in diameter is low in clean as well as polluted air. Even in polluted air these large particles have concentrations less than 10 per liter unless the sampling is done close to a source such as a dusty road. In the smaller size range of visible particles (close to 0.1 micron), an appreciable number, ranging from 50 to 100 are found in clean air; in polluted air the concentrate is 50 times higher. These higher concentrations of visible particles in city and industrial air are responsible for the restricted visibility (the industrial haze), the weight of the dust "fallout," and the "soiling index" that affects the appearance of clothes, buildings, and windows. The largest of these particles make up the dust along the city streets.

The only natural phenomena that can approach the concentration of visible manmade particles of a city come from volcanic eruptions, large forest fires, and massive dust storms. These are cataclysmic and are thus not continuing sources. Table 1 (p. 15) lists particle concentrations on the global scale.

To measure the number of fine particles in the air, a Gardner Small Particle Detector is used (produced by Gardner Associates, Carman Rd., Schenectady, N.Y.). This is a refined cloud chamber similar in principle to the Wilson-type device described in Chapter 10 (pp. 293–6) and illustrated in Fig. 36 (p. 295). The sample air is drawn into a moisture chamber; a vacuum is produced in a second chamber; the ports are closed; and a small collimated beam of

Fig. 31 Types, sizes and concentrations of airborne particles and their properties.

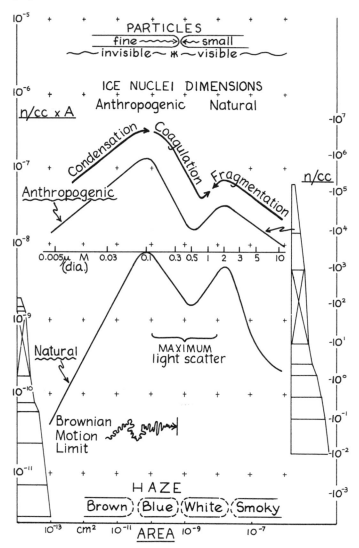

light is directed down the sample chamber to fall on a photocell.

After the light intensity is brought to maximum level, as indicated by an ohmmeter, the sample air is permitted to suddenly expand into the vacuum. As a result, a cloud forms on the particles contained in the sample air, reducing the light received by the photocell. The amount of vacuum used determines the degree of supersaturation that develops in the neighborhood of the particles. Thus a vacuum of 51 cm (20 in.) of mercury will produce a supersaturation in excess of 300%. As a result, practically all of the particles larger than 0.005 micron are forced to accept water through adiabatic expansion. The opacity of the cloud is measured by the photocell and displayed on the ohmmeter. If a considerably lower vacuum is used — 2.5 cm (1 in.) of mercury — the supersaturation is about 1% and thus measures particles of about 0.1 micron diameter. The meter reading may be applied to a calibration curve to determine the number of particles per cubic centimeter in the air sampled.

This instrument has a range for measuring concentrations as low as 100 per cubic centimeter to an excess of 1 million per cubic centimeter. It is a highly reliable, rugged, self-contained, foolproof, and easily carried instrument.

The larger particles that float in the air and produce the hazes that restrict visibility can be measured with the Royco Counter (Royco Instruments, Menlo Park, California). This measures particles having diameters of 0.5, 1, 2, 3, 5, and 10 microns. Each particle above a selected size is counted electronically as it passes through a tiny hole. The size-sensing element is based on the amount of light scattered by an illuminated particle.

The concentration of fine particles found with the Gardner Counter and the larger ones with the Royco Counter establishes the air quality of a particular region. These particle patterns have a profound effect on the amount and type of precipitation that occurs and the lifetime of the clouds that form.

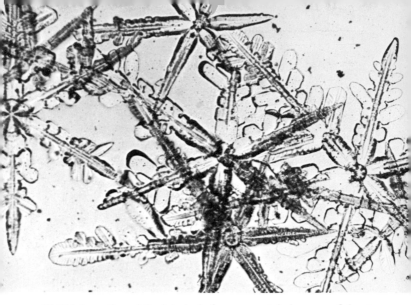

Pl.241 A portion of the interlocked snow crystals in a snowflake. As many as 50 crystals may join to form such particles. They generally contain a single crystal type.

220

Pl.242 Submicroscopic silver or lead iodide particles serve as very effective nuclei for ice crystal formation; these complex and beautiful hexagonal ice crystals form on them.

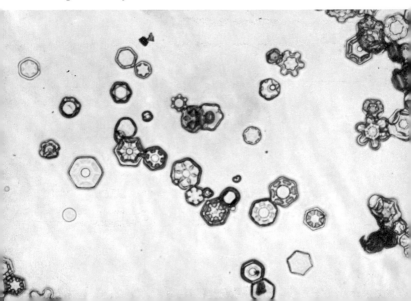

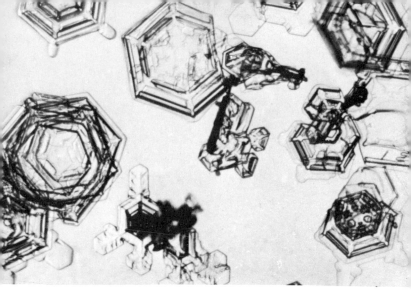

Pl.243 Crystals observed during a fall of small hexagonal plates. Notice the many interesting departures from crystalline symmetry that appear in this random collection that fell from the sky.

Pl.244 A group of capped columns, snow crystals that have formed in two different temperature regions. The crystals started growth as hexagonal columns, then shifted to plate growth.

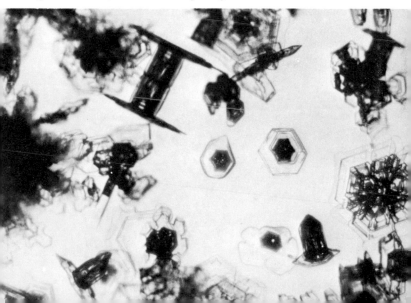

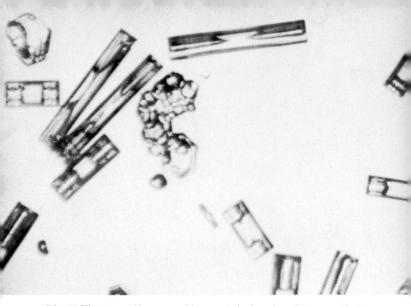

Pl.245 The type of hexagonal ice crystal often found in snow that has fallen from warmer air spread above a mass of very cold air located at ground level. Amount of snowfall is slight.

222

Pl.246 A mixture of snow needles, sleet, and rain indicates a very complex weather system. Ice needles falling into rain-forming clouds collide with raindrops, some of which freeze to form sleet.

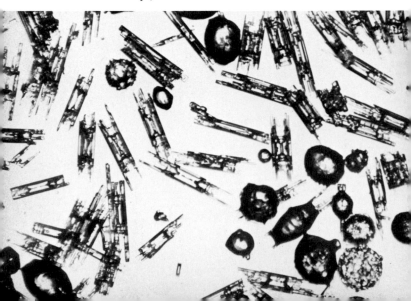

Pl.247 An assemblage of snow crystal photomicrographs illustrating the fidelity of replica technique in preserving surface detail. Plastic replicas are easily prepared.

Pl.248 Typical plastic replica of stellar snow crystal captured on a piece of black velvet. It was then transferred to a drop of replica solution on a glass slide. See pp. 288–90.

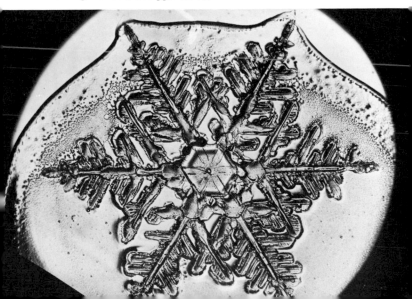

8

Weather Modification

In 1946, by a serendipitous occurrence, Dr. Schaefer discovered a highly effective method of converting a supercooled cloud to a mass of ice crystals using solid carbon dioxide (dry ice). He had been searching for a method to seed a supercooled cloud. Following the discovery of the dry ice seeding method, he established that any substance colder than $-40°C$ ($-40°F$) would spontaneously generate great quantities of ice embryos through homogeneous nucleation. These techniques are described in detail in Chapter 10.

Later that year a supercooled cloud in the natural atmosphere was converted to ice crystals, and during the next 5 years about 200 field experiments were carried out to explore the possibilities and limitations of cloud seeding.

In the same year, Dr. Bernard Vonnegut, a colleague of Dr. Schaefer's, found that silver iodide could be used in a similar manner. Since this substance could also be used outside of clouds, it has remained the major seeding agent in weather modification activities, although the use of dry ice is again becoming important.

Following these discoveries, Project Cirrus, a combined government and industrial effort, was organized in 1947 to explore the possibilities and limitations of weather modification. For the first 20 years that the seeding of clouds was conducted, there was much controversy about the reality of its effectiveness in the free atmosphere. It is now generally agreed that it is quite feasible as a way to disperse supercooled fog at airports, to carve holes in supercooled stratus decks, to change the electrical nature of thunderstorms, to increase convection in large cumulus clouds, to reduce hail damage, to increase the snow pack in mountains, and to increase rainfall by 10–30% where there are substantial amounts of supercooled clouds. Whether larger storms like tornadoes, hurricanes, typhoons, and large cyclonic storms can be affected in a useful manner requires much more field knowledge and better forecasting techniques.

Whenever natural clouds supercool, the possibility exists that the clouds can be rapidly modified, since supercooling is a highly unstable condition. Most of the larger clouds that form in the earth's atmosphere go through a supercooled phase. The amount and duration of the supercooled phase depends on the concentration of naturally occurring ice nuclei and the dynamics of the weather system. There is no question that a supercooled cloud can be completely converted to ice crystals by the use of proper seeding techniques.

There is also agreement that supercooled fog at airports can be dissipated, and such operations are now routine at major air fields throughout the northern latitudes.

Strong evidence is emerging that severe damage from hail can be reduced or even eliminated, and that the electrical structure of thunderstorms can be modified, with the possibility that the number of fire-inducing lightning strikes could be reduced or prevented if desired. There is general agreement that precipitation can be augmented by at least 10% and if optimum conditions occur the increases may be 2 or 3 times greater.

Project Cirrus demonstrated that a solid deck of supercooled clouds could have holes cut into it and that, if desirable, the entire overcast could be transformed to light snow showers. With the importance of solar energy for the heating of homes and other buildings, for refrigeration and related processes, the intentional wide-scale removal of thin supercooled stratus clouds that effectively prevent the sun from activating solar panels and heat collectors is likely to become of considerable economic, social, and engineering importance.

The scientific principles of cloud and weather modification are based on the presence (or absence) of microscopic particles in the atmosphere that act as nuclei for ice crystal formation when they enter the supercooled regions of a cloud.

Measurements on the worldwide scale show that in the naturally "clean" atmosphere far removed from man's activities, the number of ice nuclei ranges from 0.1 to 10 per liter. This concentration is 100 to 1000 times lower than that of cloud condensation nuclei under the same conditions.

It can be shown experimentally that in clean-air areas the concentration of ice nuclei can be increased by 100 times the average value without overseeding the clouds. If, however, the ice nuclei concentration becomes 10 times more, this does happen.

Overseeding is a condition that occurs when ice crystals that form are so small and so numerous that they cannot grow large enough to have an appreciable falling velocity. With overseeding, a stable zone of "floating" ice crystals develops. It is this condition that produces a brilliant undersun, a striking phenomenon frequently also seen under other conditions downwind of large cities in the wintertime. This appears to represent a classic example of inadvertent cloud modification.

In weather engineering the attempt is made to introduce from 10 to 50 per liter of nuclei for ice crystal formation effective at a temperature of $-10°C$ ($14°F$) or warmer.

As noted, the original material used by Dr. Schaefer for seeding supercooled clouds was crushed dry ice. When placed in moist air, this material produces ice embryos in fantastic numbers through the phenomenon of homogeneous nucleation. This is a process whereby any substance or material colder than $-40°C$ ($-40°F$) produces the same effect as occurs with dry ice. Dry ice, however, is the least expensive and most versatile material for cloud seeding purposes since it is effective at all temperatures colder than freezing. A fragment the size of a pea will fall through the air for about a mile before it completely evaporates.

The disadvantage of using dry ice for cloud seeding is that, to be effective, it must be put into air that either contains a supercooled cloud or is supersaturated with ice (for a more complete explanation of the latter, see pp. 71–9). Also, it is generally necessary to use airplanes for conducting dry-ice seeding, although shells and rockets can be used.

Since 1953 silver iodide has been the preferred substance for cloud-seeding activities, since it can be released into warm, dry air and will remain as an effective nucleus for ice crystal formation unless deactivated by exposure to ozone or some other agent. One limitation of silver iodide is that it is not active at temperatures warmer than about $-5°C$ ($23°F$). (Lead iodide may be substituted for silver iodide and is active in a similar manner.)

Although a few materials other than dry ice and silver or lead iodide have been tried as supercooled cloud-seeding agents, they are not as convenient or effective. Cupric sulfide, phloroglucinol, metaldehyde, and pentaerythritol are some of these substances. Sodium chloride and other soluble salts, and even water, have been tried on clouds that are not supercooled for weather modification.

Cloud modification, as opposed to weather modification, is a natural process. In many instances the modification that occurs does not result in an efficient conversion of the cloud droplets to precipitation, but often leads to conditions that, in human terms, are wasteful or undesirable. Cloud formation requires a substantial amount of solar energy: the sun heats the earth, produces convection, and as the rising moist air cools, it reaches the saturation level and condenses on cloud condensation nuclei.

If a cloud forms and then evaporates without producing precipitation, it can be argued that a certain amount of potentially available moisture has been lost that might have produced snow, hail, or rain on the ground. It is not quite that simple.

By the very process of forming and evaporating the water in a cumulus cloud, moisture has been transported from ground level to a higher altitude in the atmosphere. If this happens to a large number of clouds, a substantial region of the atmosphere that was dry has been moistened. At some later time, and probably in a different geographic area, this moisture will serve to make clouds that are larger and more likely to produce precipitation.

In their early formative stages clouds are white and opaque, and often display interesting optical effects when they are in line with the sun or moon. At all temperatures between 0°C (32°F) and −40°C (−40°F) most cloud droplets remain liquid, i.e., they do not freeze even though the temperature drops below the so-called freezing point. Thus cumulus clouds in particular show no change in appearance as they pass through the 0°C level and become supercooled. When, for example, the sky is dotted with small cumulus clouds and they are viewed at a distance, the individual clouds appear to have many features in common: They have flat bases at a certain altitude; their growing tops have firm outlines; and their sides may have a wispy appearance.

If a specific cumulus cloud is selected for continuing observation, it is likely that it will go through its life cycle in 15–20 minutes. The cloud appears first as a tiny cloudlet, then rapidly expands as it develops a flat base that marks the saturation level (100% relative humidity) of moist air rising in a convective column from the heated earth. Its upper surface looks very much like a cauliflower, marking the edge of the rising turbulent cloud-filled air. If there is also some horizontal wind shear, the cloud may assume a wedge-shaped profile with small wavelets forming and curling over like the breaking waves in water.

In 10 or 15 minutes the top of the cloud changes appearance. The previously sharp edges become ragged and wispy, and the top flattens. Often within 5 minutes more the cloud disappears entirely through the evaporation of its droplets as the air loses its buoyancy and sinks.

Quite often nothing visible remains where the cloud existed. At other times, as the cloud passes through its maximum growth stage, its sharp edges will become diffuse and more persistent and visible. Compared to the wispy fragments that leave the sides of the cloud during its formative stages and disappear in less than a minute, the fuzzy smokelike areas are more persistent and remain for some time after the cloud has disappeared. In most instances, except in the trade-wind regions of the world's oceans, this residue consists of ice crystals and the cloud is said to have glaciated. Some of the illustrations show cloud residues of this type. At other times when the cloud disappears, it leaves behind a visible region of mist or drizzle droplets considerably larger than the original cloud droplets. This residue, unlike that of a glaciated cloud, exhibits a misty appearance and the drops are large enough to have an appreciable terminal velocity. They then fall toward the earth in small shafts of precipitation.

As mentioned earlier, such effects are rarely seen except in oceanic areas where, with few cloud condensation nuclei present, the coalescence process is quite efficient and the trade-wind clouds rarely grow high enough into the sky to reach the freezing level, since their vertical growth is generally limited by the trade-wind inversion that exists at 1.8–2.3 km (6000–7500 ft.) above the sea.

With the much commoner phenomenon that can be seen over the continents in the glaciation of clouds, the ice particles are often so small that they either remain "floating" or fall slowly toward the ground as filmy streamers called virga.

These residues of glaciation or misty rain are good examples of cloud modification. Small liquid droplets with a diameter of perhaps 12 microns have either coalesced with their neighbors to form larger droplets or ice fragments, or nuclei have caused the cloud droplets to evaporate, with their water molecules streaming toward the ice. This phenomenon of vapor transfer from liquid to ice is the basis of most cloud modification processes.

Whether the change of phase was due to natural processes, to some overt action on the part of a weather engineer, or to a plume of air pollution does not alter the fact that the cloud was modified. When this occurs on an extensive scale, the effect may lead to weather or even climate modification through a change in the precipitation pattern.

The photographs preceding this section that illustrate precipitation effects are mostly examples of natural cloud modification. Most of those that follow show instances in which intentional or inadvertent manmade modification has occurred.

While arguments among atmospheric scientists still go on concerning the degree to which intentional efforts to modify clouds can augment, redistribute, or prevent precipitation, the argument about whether or not supercooled clouds can be modified by seeding is no longer heard.

Whenever natural mechanisms fail to produce a substantial conversion of the cloud droplets to precipitation that reaches the ground, or if the results lead to such things as damaging hail, fire-producing lightning, damaging wind, excessive precipitation, 100-mile-long streamers of false cirrus, supercooled ground fogs, or extensive sun-blocking supercooled cloud decks, there is the possibility that intelligent weather engineering can produce more desirable situations (from the human standpoint) that are also ecologically acceptable.

All of these phenomena are part of the global ecology, some aspects of which are just beginning to be understood. Plans for intentional cloud modification should not overlook potential alterations that may occur in the ecosystem if such efforts are successful.

Pl.249 The Mt. Washington Observatory in the White Mts. of New Hampshire. Studies of supercooled clouds during the winters of 1943–45 led to discovery of dry-ice seeding by Dr. Schaefer 1946.

Pl.250 The first cloud seeded by Dr. Schaefer 11/13/46 using 6 lbs. of crushed dry ice. It was an altocumulus at −18°C (−1.5°F) located near Greylock Mt. in w. Massachusetts.

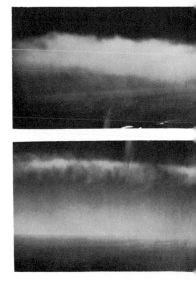

Upper left: Southern end of the 6.4-km (4-mi.) cloud before seeding (10:34 A.M.); base at 4.2 km (13,600 ft.). *Upper right:* The same section of the cloud about 5 minutes after seeding. Note the long "draperies" of the sun. *Lower right:* The midsection of the cloud about 15 minutes after seeding.

Pl.251 Condensation trails formed in moist cold air when tiny grains of dry ice fall through it. Trails comprise myriads of ice embryos generated by localized chilling of air below −40°C.

230

Pl.252 Featureless hexagonal plates that form on ice embryos generated by homogeneous nucleation, which occurs when moist air is colder than −40°C, as at top of troposphere.

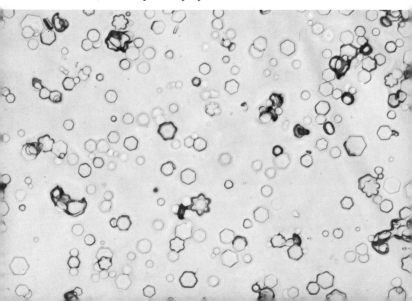

Pl.253 Triangular trail of growing ice crystals produced in supercooled cloud deck by a B-17 aircraft dropping crushed dry ice as it flies close to upper surface of cloud. Project Cirrus, 1947.

231

Pl.254 Long rectangular holes cut through a stratus cloud using crushed dry ice at rate of about 1 kg/km. Such holes develop in less than an hour and become at least 2 km wide. Project Cirrus, 1947.

Pl.255 One of the first seeding runs of Project Cirrus used crushed dry ice at rate of 0.45 kg/km along L-shaped path. View 33 minutes after seeding; dark area with undersun is ice crystals.

Pl.256 Sheet of ice crystals produced at about 6 km altitude by dropping crushed dry ice from an airplane. Presence of moist air supersaturated over ice causes ice embryos to grow.

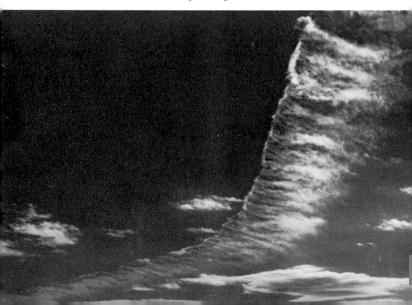

Pl.257 Effect produced in 24 minutes after dropping 34 kg (75 lb.) of crushed dry ice into top of clouds supercooled to −5.6°C (22°F). Length of a long leg of race track pattern 27.4 km (17 mi.).

233

Pl.258 A large cumulus cloud seeded by airplane with small fragments of dry ice. The seeded cumulus towers merged as rain started to fall east of the Manzano Mts. in New Mexico.

Pl.259 Two lines cut through a deck of supercooled clouds, using dry ice fragments dispensed at rate of about 1 kg/km. Thin veils of ice crystals remain in seeded area but most have fallen.

234

Pl.260 An extensive hole cut through cloud deck shown above. This opening developed in about 40 minutes and remained open for several hours. New clouds are starting to form in cleared area.

Pl.261 A solid deck of supercooled clouds seeded with burning pellets of charcoal containing silver iodide. The nuclei in the smoke converted mile-wide strips of cloud to ice crystals.

235

Pl.262 The same area as shown on Plate 261, from the other end of the seeded field about 10 minutes later as the ice crystals settled out of the cloud to produce lines of virga below cloud base.

Pl.263 First cloud seeded at a distance by a ground-based silver iodide generator located about 32 km (20 mi.) upwind. Clouds formed near Albany, N.Y., with generator in Schoharie Valley.

236

Pl.264 Three small supercooled cumulus clouds that formed over valley of Lolo Creek in Bitterroot Mts. of Montana. Middle cloud seeded with silver iodide and completely glaciated.

Pl.265 Localized ice crystal cloud produced by throwing water into the air when temperatures are colder than −40°C. The water at such temperatures produces ice crystals spontaneously.

Pl.266 A cloud produced similarly to one on Plate 265 at a temperature of −44°C (−47°F) beside Old Faithful Geyser. A dense ice fog also formed after each geyser eruption, through the same mechanism.

Pl.267 Billions of ice crystals being generated in a supercooled fog by swinging a wire basket containing chunks of dry ice. These modify the winter fogs at the Old Faithful Basin at Yellowstone.

238

Pl.268 A sunstreak and 22° parhelia produced by ice crystals formed in a supercooled fog that was seeded with dry ice. Such fogs form frequently during winter at Yellowstone Park.

Pl.269 A cumulus cloud seeded with 1 kg/km of crushed dry ice dropped into it early in its growth cycle. The cloud system over Rio Salado in New Mexico rained for 4 hours.

Pl.270 A high tower of cumulus cloud (like that shown on Plate 269) formed rapidly after silver iodide was put into a similar cloud mass during same seeding operation over Rio Salado.

Pl.271 Industrial smoke producing a massive volume of gases and particles. Concentration of cloud condensation and ice nuclei produced in this manner modifies natural processes.

Pl.272 Snow showers falling from clouds downwind of the large industrial complex shown on Plate 271. A good example of precipitation that has likely been modified by the pollution plume.

Pl.273 Ice nuclei convert a supercooled plume from a large steel plant to ice crystals. A similar water droplet cloud from the oil refinery at left remains unaffected.

241

Pl.274 A dissipation trail produced in supercooled cirrocumulus either by ice crystals from condensation trail of an airplane or by heat of exhaust gases. The former is more likely.

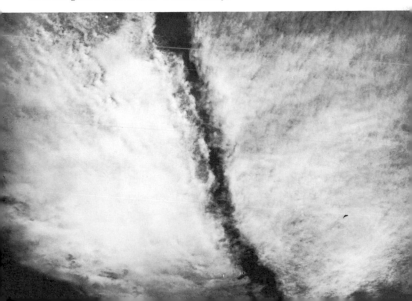

9

Precipitation — Forms and Effects of Ice and Water

Much of the world's precipitation begins as ice crystals or frozen water in the form of hail, ice pellets, snow grains, graupel, or sleet. Entering warmer air, the smaller frozen particles often melt before they reach the ground, although hailstones frequently reach it even at hot temperatures. In some parts of the world, especially in the latitudes of 10°–30° north and south (where the trade winds blow), precipitation may form in clouds that never cool to 0°C (32°F). Such rains generally occur in the maritime areas where there are minimum numbers of cloud condensation nuclei (generally less than 100/cc). Since relatively few cloud nuclei are present in ocean air and they have varied sizes, the cloud droplets grow rapidly by collision and coalescence to form rain. When major weather systems such as typhoons (hurricanes) or big cyclonic storms sweep across these areas, they grow so large and high that supercooled clouds form and ice crystals control the precipitation process.

This chapter illustrates some of the more commonly occurring forms of precipitation, with the major emphasis on solid forms since they are most readily observed. The variety of such particles is quite remarkable; the chart of frozen precipitation shown in Appendix 1 is greatly simplified, and each of the 10 categories could be further divided into 10 or more subforms. Thus precipitation resulting from the wide variety of interacting causes that occur in cloudy air can be classified into 100 or more categories. The examples shown, however, will be sufficient for all except the specialist.

Snow and Ice Formations

Many users of this guide may live in regions where snow rarely if ever reaches the ground and the temperature never drops to the freezing point. Even in these areas, however, snow can be seen by looking up at the sky: cirrus clouds are composed of ice crystals, and the upper portions of shafts of heavy precipitation generally consist of either snow, graupel, or hail.

Since water in a solid state assumes such interesting structures, some of them are shown in this guide. These illustrations will be useful for identifying forms of precipitation that may be difficult to recognize.

Some of these structures develop simply by the action of the wind or fallen snow. Others are the result of interactions between wind, temperature, liquid water, and water vapor, separately or in sequence, often under unusual circumstances.

As with the myriad forms that snow crystals assume, the structures illustrated in the following pages may be expected to occur in bewildering variety in response to variations in temperature, wind, moisture supply, and other factors that combine to make them form and grow. Some, like foam volcanoes, are evanescent; others, like ice stalagmites, may develop throughout the entire winter, serving as a fascinating, complex record of liquid water supply, temperature, dissolved gases, and sometimes air motion. The intriguing feature of these latter formations is the challenge they present to the observer who tries to learn the sequence of events that led to their formation.

Liquid Water Precipitation

Mist In a stratus cloud the droplets have varying sizes depending on the number of cloud condensation nuclei in the air. If less than 100/cc are present in the initial cloud, droplets that form by coalescence are often large enough to be felt on the face of an observer standing in air moving 1 meter/second. Similar particles may form over the sea or over the land and lakes in clean air. Mist may also form in stratus clouds having 1000 or more cloud condensation nuclei per cubic centimeter. Particles of this type grow very slowly, and when some of them reach a size sufficiently large to precipitate, they fall in a concentration about 10 times lower than the mist that forms in cleaner air.

Oregon or Scotch Mist, as well as the so-called "black" stratus often seen in trade-wind regions, are terms used to describe the occurrences of misty precipitation. Mist particles range in size from 50 to 500 microns.

Drizzle Composed of uniform small raindrops varying in size from .2 to .5 mm, drizzle falls out of stratus clouds from several hours to a day or more. This type of precipitation causes very little erosion and soaks into the ground quite effectively. Rainfall amounts range from 0.1 to 0.5 cm per hour and the cloud water in which this precipitation forms is 0.1–0.3 g per m^3.

Rain Raindrops measure from about .2 mm to as large as 1 cm in diameter, but most frequently are 3–6 mm. Rain may fall from nimbostratus or stratus-cumulus combinations, including such giant storms as large thunder- and hail-storms, tornadoes, hurricanes, and cyclonic frontal systems. This type of precipitation generally amounts to 0.5–2.5 cm an hour, though it may be several times greater. When the rain cloud is cumuliform, the amount that falls often shows high variability over the region receiving the rain. The clouds in which this type of precipitation forms have a liquid-water concentration of 0.5–5 g per m^3.

Freezing Mist These particles have a size range similar to that of warm mist, but they become supercooled as they fall through cold air. Freezing mist may cause more ice damage than freezing rain. If it lands on a tree limb, little will dribble away; 5–10 cm (2–4 in.) of clear ice may form, causing massive damage. In an urban area, this may paralyze telephone and power services for long periods; typical damage is shown on Plate 305.

Freezing Rain and Sleet The rattle of sleet on windowpanes is a characteristic sound during a freezing rain. Frequently the beginning of this type of storm involves a fall of these frozen particles followed by a drizzle-rain that slowly warms as the overriding warm air displaces the colder air near the ground. Sleet particles often have a curious structure. As the drizzle-rain falls into the cold air, it freezes from the outside inward. Just before the drop is completely frozen, the pressurized water trapped in the center by the freezing mode cracks the ice sphere, flows to the surface, and upon emerging, freezes in the form of a spicule. Other partly frozen drops may break open when they hit the ground, leaving cuplike fragments, or some other drops may combine with ice crystals or transform them to lumpy, transparent ice fragments; strange crystalline spheroids will occasionally be found mixed with the more common frozen droplets. A storm that produces freezing rain has a warming trend and often ends before serious damage occurs. The precipitation size range is similar to that of drizzle.

Solid Water Precipitation

Graupel Sometimes called soft hail or snow grains, graupel consists primarily of a mass of frozen cloud droplets. Sometimes the initial particle consists of a cluster of ice needles or a stellar crystal; when this happens the particle tends to be conical in form. If the initial particle is primarily an aggregation of cloud droplets, the graupel appears as a lumpy, somewhat spherical particle. These particles are soft, and often when they hit an object they flatten out to appear as a rounded spot of powdery snow without any apparent crystalline structure.

Graupel often forms in strong updrafts of cold air and in blizzards and "lake effect" snowstorms. The particles are often highly electrified, and each will produce a static noise upon impact with a radio antenna, either while frozen or recently melted. Graupel is often found in severe lightning storms and frequently serves as the core of a hailstone. The particles are 1–7 mm in diameter.

Hail This, of all the forms of precipitation, produces the greatest immediate damage. Hail particles form in the core of large convective clouds of the cumulonimbus type and may grow to maximum size within a small region of an intensive updraft, which is called the accumulation zone and has a liquid-and-ice-crystal content of 10 g per m³ or more. Within this zone the particle is in a balanced mode — its increasing weight is counteracted by an ever-increasing upward air velocity. At times a hailstone may start falling only to encounter a new and more intense updraft that will carry it even higher into the cloud system and perhaps toss it out at the top, so that it falls to the earth through clear air. At other times it may fall through the cloud as the updraft weakens, sweeping up more cloud droplets and ice crystals until it leaves the base of the cloud. Small hailstones that have melted while falling may have a diameter of 10 mm or more and will produce splash marks 3 cm or more in diameter, often containing air bubbles. The cross section of a large hailstone commonly appears to have an onionlike structure of transparent and translucent layers. Polarization studies show the "stone" to consist of crystalline units ranging in size upward from 10 to 2000 microns but at times with structural units as large as 20 mm. These units indicate periods of wet and dry growth; the larger crystals form during wet growth and the smaller elements represent frozen cloud droplets and ice crystals cemented together by moisture during dry growth. The translucent ice is caused by tiny air bubbles. These can be isolated by melting the hailstone in warm glycerine. When it is melted, a 10-mm translucent hailstone will sometimes be found to possess 10 million air bubbles!

The damage caused by hail is often intensified by the accompanying wind produced by the spreading out of the cold downdraft that the falling hail column generates. Hailstones can be 3 mm to 6 cm or larger.

Snow The crystalline nature of snow allows it to assume many forms. These reflect the temperature and moisture conditions in the supercooled clouds in which the particles form. The relationships between these factors were established in 1936 by the classic work of Ukichira Nakaya in the snow country of Hokkaido, Japan. Appendix 1 was prepared by Dr. Schaefer in 1952 with the personal help of both Professor Nakaya and Dr. Marcel de Quervain of Switzerland. This chart includes all of the basic forms likely to be found in solid precipitation. A full description of the causes of these crystalline patterns is beyond the scope of this *Field Guide,* but literature is cited in the *Bibliography* for those wishing further information. The descriptions that follow provide a general summary of the varieties that occur.

The density of freshly fallen snow tends to be about 0.1 because of the loose packing that initially occurs. The wetter and warmer varieties of frozen precipitation, especially sleet and hail, may have an initial density as high as 0.5 (half as dense as water). All snow tends to approach a density of 0.25 when it consolidates by sublimation, localized melting, and recrystallization, and is called "ripe" snow by the hydrologist and "corn" snow by the skier. The particles become spheroidal in shape and loose, with some degree of cohesion due to the water film normally present during the day. An exception to this occurs in the formation of "sugar snow" or depth hoar. This is a major recrystallization that begins to form under deep snow at ground level and proceeds upward. It is caused by a microcirculation of air around the grains in a porous snow pack; the old snow evaporates and the vapor water molecules condense in the form of cup scrolls and other asymmetric types of clear ice crystals. Crystals of this kind grow in detached form so that there is no adhesive strength to the underlying layer and the snow flows like sugar. Weak layers of this snow are the major cause of slab avalanches.

The lowest density of snow occurs when very large and perfectly symmetrical stellar crystals fall in calm air with a temperature of about $-15°C$ (5°F). Since large stellar crystals tend to tumble as they fall, their orientation on exposed surfaces is random, with the tips of the crystals interlocked. Densities as low as 0.003 have been measured in snow depths of 60 cm (2 ft.). The slightest disturbance of this type of snow causes it to collapse; skiers have maneuvered waist-deep in it.

Symmetrical Snow Crystals Of the 10 categories included in the International Snow Classification, 5 may be listed as showing crystalline symmetry. The most beautiful and aesthetically pleasing forms are the plates and stellars. Fortunately these also are the forms that may be encountered by the dedicated observer and are the crystals that can most easily be formed in the laboratory or home cold chamber (see pp. 280–82).

Snowfalls of perfect crystals of these 2 types, especially the stellars, are relatively rare except under special conditions. Entire snowstorms lasting many hours frequently contain few, if any, perfect crystals. At times, however, especially in quiet air at night or shortly after dawn, practically all of the crystals that fall have perfect symmetry and are worthy of replication or photographic documentation as described in Chapter 10. While aesthetically pleasing, such perfect crystals are not representative of the average snow precipitation.

The other 3 symmetrical crystals (the columns, needles, and capped columns) are more frequently found in high concentrations during extended storms. Needles occur under the warmest conditions of crystalline symmetrical snow development, and generally fall with no other varieties present except occasional graupel and sleet. As with graupel and sleet, considerable atmospheric electricity accompanies heavy falls of needles, sometimes culminating in lightning. Needles may produce snow accumulations in the range of 1–2 cm per hour.

Columns and capped columns generally occur at temperatures colder than $-10°C$ (12°F), often preceded by a dramatic thickening of cirrostratus clouds. The 22° halo and snow sun are precursors to a fall of these crystals. Sometimes hexagonal plates are found among the hexagonal columns of cirrus storms, but they are thicker than those that form at lower levels. Since cirrus clouds contain very low amounts of moisture, such crystals are very small and the amount of accumulated snow is rarely greater than 1–2 mm per hour. Capped columns include some of the most remarkable and bizarre forms of ice crystals. Some of them may combine as many as 3 different crystalline forms — the column, plate, and needle — reflecting in a dramatic way the very cold (−40°C or colder) environment in which the columnar crystal originated, the intermediate temperature range of −10° to −16°C in which the capping of the column by hexagonal plates occurred, and the still warmer clouds (between 0° and −8°C) in which the needles developed.

On the average, the capped column crystals are considerably larger than the columns. The snow accumulation of 0.3–1 cm per hour reflects this increased moisture availability.

Asymmetrical Snow Crystals The less interesting types of snow crystals are the spatial dendrites and irregular crystals. These 2 forms are the most plentiful.

Spatial dendrites are complex, somewhat irregular shapes of stellar- or plate-like crystals that form in a 3-dimensional rather than a flat array. They generally occur with an abundance of cloud moisture available and in high concentrations. Snowfalls augmented by the lifting of air over mountains and the presence of multiple cloud layers seem to favor the occurrence of these types of crystals. Their size is 0.1–0.5 cm in cross section.

On very cold clear mornings, shortly after the sun has risen, there will often be a fall of ice prisms descriptively referred to as diamond dust. They frequently provide a spectacular sunstreak or sun pillar. Despite their high visibility due to the sun reflection, they are generally microscopic. These crystals are platelike but rarely symmetrical. Quite often they have 3 or 4 dominant prismatic arms, the others being small or invisible. They generally consist of clear ice with many flat surfaces joined together by ridges. They occur in such low concentration and for such short periods of time that they never provide any measurable accumulation. The size range is 0.1–0.5 mm in cross section.

Rime When supercooled clouds sweep over a mountain summit, they often produce substantial deposits of rime ice feathers, formations that grow into the wind on most exposed objects as the cloud droplets freeze upon impact. These are of varying density and appearance, depending on temperature, wind velocity, cloud drop size, and turbulence. Since they are formed by the freezing of cloud droplets, their structure may be very grainy if the drops freeze instantly, or very smooth if the cloud water has a chance to flow before it solidifies. Rime formed by flowing water often contains

large crystals. The higher the wind and the larger the supercooled
cloud droplet, the larger and more rapid the rime formation. There
are many forms of rime, ranging from icy to powdery, and including solid, granular, and feathery. Structures up to a meter in
length commonly form on trees and other objects on mountain
summits.

Other Effects Produced by Ice and Water

In addition to the unusual forms of ice that occur under special
circumstances such as foam volcanoes, hydromites, and sastrugi,
there are secondary effects that are equally interesting. Some are
the result of water action; others arise from the movement of ice.
To name typical examples: the formation of varve deposits in
post-glacial lakes; ancient ripple marks in rocks that once were
sand in shallow seas; mud cracks in evaporated ponds; scratches
from rocks at the base of glaciers; trees shattered by cataclysmic
floods and ice storms; and the deposition of frost on spider webs.
Other effects illustrated in this chapter relate to geysers, hot
springs, gas bubble formation in ice, the creep of snow under the
effect of gravity, and a variety of related phenomena.

There are many other fascinating things that can be seen, some
of which are so evanescent as to defy successful photography,
while others require such an unusual combination of conditions
that they are extremely rare. Time spent in searching, observing,
and recording such effects is highly satisfying.

Pl.275 The near juncture of a huge icicle and an ice cone, which will soon form a pillar. Similar structures develop in caverns when a stalactite merges with a stalagmite.

Pl.276 Massive ice structures that have grown on small tree twigs located near the base of a waterfall. The spray droplets freeze upon contact, growing toward the source of the spray.

Pl.277 A "frost flower." This ice formation is similar in structural growth to columnar frost. A few plants, notably Dittany and Frostweed, show similar growth patterns.

250

Pl.278 Moisture oozing from soil when the air temperature is very cold freezes to form hairlike or dense masses of columnar frost. The growing portion of these crystals is at or under the soil surface.

Pl.279 Hoarfrost fernlike ice crystals grown on grass blades at the opening of a tiny cavern from which moist air emerges into the wintry cold. Such formations are extremely fragile.

Pl.280 Round deposit of sticky snow formed on telephone wire. Gradual turning of the wire as it becomes coated with snow accounts for the uniform deposition layer.

Pl.281 Foam volcanoes form below waterfalls on ice-locked streams when a thaw is followed by a cold, calm night. If bubbles form in the water, froth oozes from holes and freezes as hollow cylinders.

Pl.282 Snow deposits on reeds and other growths in the run-off channel of a hot spring. The interlocking of snow crystals produces many strange and beautiful formations.

Pl.283 Hydromites form in cold air near the mouth of caves when water drops fall from the roof and freeze. Their growth depends on the water supply, the temperature, dissolved gases, and airflow.

Pl.284 Dr. Schaefer with a student in the Tory Hole in the Helderbergs of eastern New York. These hydromites are similar in growth mode to the much slower formation of stalagmites in caves.

Pl.285 Snow "igloos" form when a heavy snow covers warm ground, as in the geyser areas of Yellowstone Park. After the storm, heat from the earth causes the snow to shrink in this way.

254

Pl.286 Unusual type of snow igloo with central hole. No obvious reason could be discovered to account for aperture. However, possibly enough vapor was rising from soil to form opening.

Pl.287 Strong wind following a fresh snowfall removes some of the snow to disclose variations in crystal type, adhesion, and density of fallen snow. These erosion patterns are quite beautiful.

Pl.288 Replicas of snow on ground. A 40-cm (16-in.) snowfall was sampled at 24-hour intervals as snow became consolidated by recrystallization. Additional snow and sleet fell on sixth day.

Pl.289 While sastrugi are encountered most often in windy polar regions, they also form wherever strong winds blowing snow and cold air occur. Somewhat like a sand dune slowly moving downwind.

Pl.290 A snowstorm often contains particles varying widely in size and type. If strong winds erode the snow, a striking view of these differences is provided.

Pl.291 Some ghost trees are formed entirely by supercooled water droplets freezing upon impact. Unlike rime feathers, which form under windy conditions, some show little effect of air movement.

Pl.292 The sky behind these ghost trees is completely free of light-scattering particles, unlike those shown on Plate 291, where ice crystals and water-droplet clouds provide a much brighter background.

Pl.293 The track of a "snow devil." As the whirlwind passes across a field of snow, the base of its vortex lifts thin sheets of snow from the surface to form these peculiar tracks.

258

Pl.294 Snow driven by blizzard winds builds up on trees and other objects with very high density. Such particles are very small and sometimes liquefy briefly upon impact, thus forming an icy layer.

Pl.295 A rime of mud coating shrubs near base of Grand Falls on the Little Colorado R. in Arizona. The tiny mud balls shown on Plate 296 strike the vegetation to cover it like rime ice.

259

Pl.296 Mud balls with a diameter of about 25 microns formed by the evaporation of spray from the base of Grand Falls on the Little Colorado R. in Arizona.

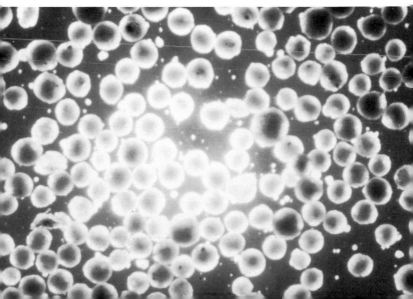

Pl.297 Spectacular eruption of boiling water and vapor into subzero air at Riverside Geyser on Firehole R. in Yellowstone. Such eruptions produce extensive supercooled fogs in the area.

Pl.298 Eruption of Old Faithful in Yellowstone when the air temperature was −44°C (−47°F). When droplets from condensed steam cool, a thick ice fog forms. No supercooling occurs colder than −40°C.

Pl.299 Hot springs and geyser runoff plumes along Firehole R. in Yellowstone Park. At night in winter such condensed steam forms a thick supercooled fog limited to 100 m (300 ft.) by an inversion.

Pl.300 The breaking of the night inversion fog at Norris Geyser Basin in Yellowstone. The heat of the sun generates convection that mixes the warmer air above with the colder air near the ground.

Pl.301 When newly fallen snow consisting of interlocking stellar crystals is warmed, the adhesion between crystals permits it to flow, producing strange structures on sloping surfaces.

262

Pl.302 Hoarfrost coating every twig of a tree. Such formations may develop during a quiet frosty night when moist air becomes supersaturated over ice. It is not necessary to have fog.

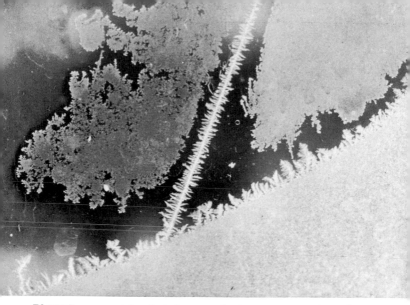

Pl.303 Large crystals of windowpane frost grow on scratches, edges, and other nuclei as the sheetlike smaller crystals evaporate. The "court" around larger crystals shows this effect.

Pl.304 The black pebble at end of this wormlike track has produced the "boring." Heated by the sun, such objects are held against unmelted ice or snow by a water film and move 2–3 mm per min.

Pl.305 When rain falls into cold air at ground level, it coats all exposed objects with an icy layer. Trees may break, power poles snap, roads become slippery, and great damage be done.

264

Pl.306 Broken treetops and branches resulting from an intense ice storm. Cold misty rain fell into ground-level air at −5°C (23°F). As ice accumulated, trees, power lines and poles were broken.

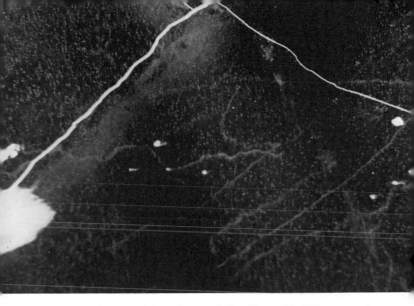

Pl.307 New ice on a lake. The wandering lines of bubbles trapped within the clear ("black") ice originate on undersurface as gas bubbles form on tiny drifting objects and are encased.

265

Pl.308 Thin sheets of ice formed at night begin melting under the sun at boundaries between individual crystals. Impurities concentrated at edges heat first and permit crystals to be separated.

Pl.309 A huge crystal of ice more than 30 cm (1 ft.) in diameter which has formed in the still waters of a pond. Such crystals may be even larger since they are started by ice nuclei in the air.

Pl.310 Myriad ice crystals from a fall of "diamond dust" seeded this newly formed layer of pond ice, resulting in such tiny crystals that they are not visible—in contrast to those on Plate 309.

Pl.311 In the wintertime an old spider web may become visible when supercooled droplets hit the web, freeze, and form strange disk-shaped ice "beads."

Pl.312 Icy, irregular-shaped "beads" on spider web formed when supercooled fog occurred. Originally deposited as droplets that then froze, as indicated by absence of crystals.

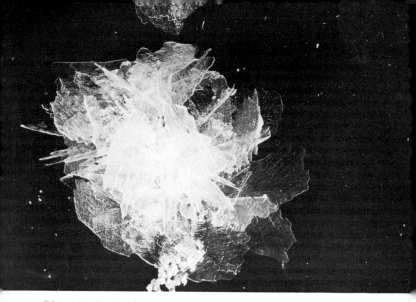

Pl.313 Anchor ice forms on underwater objects as large blades, often in single crystals. The ones in this photograph formed when a river overflowed and then receded.

Pl.314 Trash screen of hydroelectric plant plugged by accumulation of tiny disks of frazil ice, which forms in open streams in winter. Screen on left has been cleaned by hitting with sledge hammer.

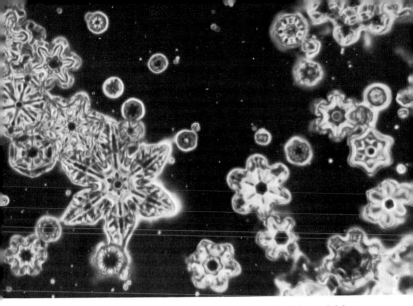

Pl.315 Ice crystals grown in the laboratory on lead-iodide nuclei from the reaction of iodine vapor with submicroscopic lead compounds from the exhaust pipe of an automobile.

Pl.316 Snow crystals found about 8 km (5 mi.) downwind of a silver-iodide smoke generator located on a mountain near Steamboat Springs, Colorado. These patterns often occur in seeded clouds.

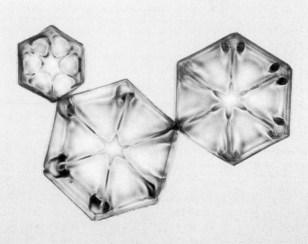

Pl.317 A typical example of the life cycle in hailstone growth shown in cross section. It formed on a graupel particle, then grew wet, dry, wet, and then mostly dry. Diameter 5.7 cm.

270

Pl.318 A symmetrical growth that started on a frozen drop (seen at upper right) after which at least seven alternations of dry and wet growth occurred. Diameter 5.3 cm.

Pl.319 Unusual growth pattern of this hailstone indicates it began on a round frozen particle containing large ice crystals, grew dry briefly, and then wet until growth ended. Diameter 5.3 cm.

Pl.320 A hailstone grown on a snow grain that in turn grew as a conglomerate of water droplets and ice particles. Early development shows dry growth. Diameter 5.1 cm.

Pl.321 Cross section of ice layers formed near a rich source of water vapor and supercooled fog droplets in subzero air above Black Sand Spring in Yellowstone. Resemble hailstones.

272

Pl.322 Banded ice formation, formed by freezing of both vapor and water, near Black Sand Spring, before being cut and polished as shown on Plate 321. Gas bubbles cause the cloudy portions.

Pl.323 Flow patterns form when water forced by internal pressure overflows from a small crater. The water carries sand grains of differing size and density to form these strange structures.

Pl.324 A thick type of hexagonal ice crystal frequently found in snow falling from cirrostratus clouds. Original center of the crystal evaporated (sublimed) during its several hours' fall.

Pl.325 Boiling water drops thrown with explosive force into the sub-zero air by Sawmill Geyser in Yellowstone, leave condensation trails when superheated water suddenly produces large bubbles of vapor.

Pl.326 Heavy frost and rime deposits covering evergreens come from supercooled fogs that occur at night along the open waters of the river.

Pl.327 Flash photograph catches the splash resulting from raindrop hitting a shallow pool. Rodlike extrusions break up into many tiny droplets that shoot up and fall in curved trajectories.

Pl.328 Splash patterns formed on a soot-coated glass slide when raindrops fell on it. Patterns develop when air is trapped under flattened water drop. Tiny bubbles and droplets lift soot.

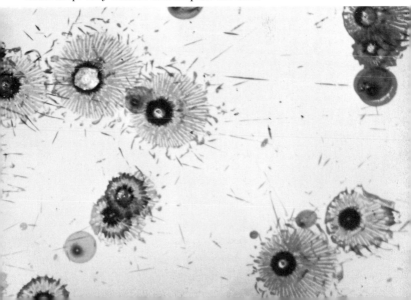

Pl.329 Langmuir Streaks modified by local water flow and terrain. Mid-channel intensification of streak, effect on wind by point of land, and lack of streaks in bay are striking.

276

Pl.330 View of an assemblage of slush ice in Lake Erie as seen from a distance of about 10 km. Streamers of slush form as strong winds produce Langmuir vortices parallel to wind.

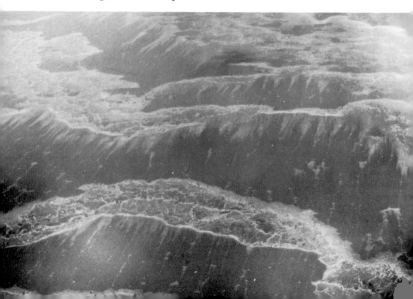

Pl.331 Glacial scratches and polishing on top of cluster of basaltic columns in a lava flow on eastern slope of the Sierra Nevada in California. Abrasive rocks were at bottom of glacier.

Pl.332 Varves are one of the telltales of postglacial weather. Finely ground rock settling to the bottom of lakes shows yearly periods, the finest particles settling when lake was ice-covered.

Pl.333 Peeling of silt deposited by water in thin layer on top of coarse sand. As the silty mud dries it contracts and then peels away from the porous sand. With a thick layer, mud cracks form.

278

Pl.334 Snow rollers form when a strong wind blows fragments of snow or ice across a flat or sloping field covered with a shallow layer of sticky snow. The roller widens as it grows. Its core may evaporate.

Pl.335 Cloud forming in supersaturated air above a hot spring along the Firehole R. in Yellowstone Park. Cloud condensation nuclei are put into the air upwind of the open pool.

Pl.336 Wing-flap condensation trail produced when moist air is expanded and cooled to form a visible vortex during rapid climb. Air in vortex may become so cold as to produce ice crystals.

10

Simple Experiments

Introduction

Not long ago the director of a large industrial research laboratory was heard to say that the days of the string-paper-and-sealing-wax method of scientific research had passed. He is right to this extent: some of us now use nylon filaments, Scotch tape, epoxy, hot plastic glue, polyethylene tubing, plywood, and a host of other simple, inexpensive, and readily available modern materials for constructing our research equipment. However, we still use paper clips, string, and sealing wax when their special features are required.

If there was ever a time when we needed bright inquiring minds, solvers of practical problems, enthusiastic, dedicated, and hard workers who are not concerned about the length of the work day and are also familiar with the technique of serendipitous discovery, it is now — in our age of the atom and outer space. No matter how large and ambitious the scientific or technological project, and no matter how well-planned and systematic it is intended to be, there will always be the necessity for quick human innovation and on-the-spot problem-solving.

It is hoped that this *Field Guide* will encourage more field research and human curiosity. The following pages will enable the reader to better understand through experiments and observations, both in the field and in home or school laboratories, some of the fascinating relationships on which the atmospheric sciences are founded. The 15 sections that follow are only a sampling of the phenomena that can be examined. They are intended to arouse both curiosity and the desire for further explorations.

I. The Mixing Cold Chamber

Among the most elegant cloud physics experiments are those that can be carried out in a simple cold chamber. The most convenient home appliance for making a chamber is a food freezer chest with a capacity in the range of 100–150 liters (4–6 cubic ft.); see Fig. 32. Larger freezers are less convenient to use.

The chamber should be lined with black nylon velvet. It is extremely important that the velvet be the semitransparent type; this transmits light and air, and permits air movement between the cold refrigerated walls of the chamber and its interior, but effectively prevents fragments of frost from entering the working area. The velvet lining may be fastened with adhesive tape to the outer edge of the chamber so that it can hang inside. A more permanent and better method is to stretch the cloth on rectangular wooden frames made of 1-cm-square strips of soft wood, one for each wall of the chamber and one for the floor.

Another important feature is proper illumination. A 6-volt flashlight, a 35mm slide projector, or a similar light should project a strong beam of light into the dark chamber from an upper corner. If a mirror is placed in the opposite corner at the bottom of the chamber, the reflected beam can be oriented parallel to the original beam or directed at another angle for further illumination.

Most home freezers of the chest type operate in the temperature range of $-20°$ to $-25°C$ ($-4°$ to $-13°F$). It is useful to have one thermometer 5 cm (2 in.) down from the top of the cold zone and another 5 cm above the bottom. These may be hung horizontally on bent paper clips hooked into the cloth.

Ordinarily such a chamber at equilibrium will possess a weak inversion with the top at $-15°C$ and the bottom at $-25°C$. At the contact zone between the room air and the cold air of the chamber,

Fig. 32 Mixing cold chamber.

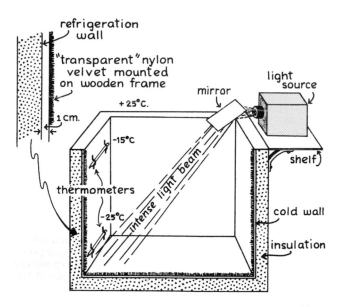

an extremely strong temperature inversion occurs. Therefore, if moisture is introduced into the cold air by placing a shallow container of 40°C (104°F) water at the bottom of the chamber, a water droplet cloud will rise toward the top of the chamber as a turbulent convective column. On reaching the top of the chamber, this cloud will surge momentarily into the warm air of the room but will then fall back to spread horizontally toward the sides of the chamber, where it will then move downward. This phenomenon is a tiny model of a large towering cumulus cloud rising into the inversion at the base of the stratosphere and then subsiding and spreading out to form the anvil top so characteristic of cumulonimbus clouds. The cloud droplets become supercooled a few seconds after rising above the warm-water source and may cool without freezing to the average temperature of the chamber — perhaps −20°C. For demonstration purposes, a supercooled cloud can be formed by slowly exhaling into the chamber or by waving a moistened cloth back and forth in it several times.

The cloud that forms in the cold chamber consists of water droplets formed by the condensation of water molecules on cloud condensation nuclei that have diffused into the chamber from the surrounding room air. The opacity of the cloud that forms depends on the number and size of the cloud droplets that have grown on the active cloud nuclei. Striking a match inside the cold chamber will cause a spectacular change as a new cloud is formed because of the increased number of active cloud condensation nuclei produced by the ignition process. The cloud becomes more noticeable due to the decreased size and greatly increased number of its droplets.

The mixing of colder air into the rising column of moist turbulent air from a warm-water source (the breath or a moist cloth) is so effective that by the time the air has moved 30 cm (12 in.) it has cooled from its initial temperature of 35°–40°C (95°–104°F) to colder than 0°C (32°F).

Such a cloud can readily be "seeded" using a variety of simple techniques. One of the easiest and most spectacular is to hold a small chunk of dry ice in a gloved hand above a stable, quiet, supercooled cloud in the chamber. Then scratch the dry ice with a sharp point so that tiny fragments fall into the cloud. A spectacular effect occurs as each dry ice particle generates a vertical condensation trail containing many thousands of tiny ice crystals. These grow rapidly and produce a chamber filled with twinkling crystals. If the air is calm, each condensation trail will quickly be surrounded by a dark zone a centimeter or more in diameter. This is caused by the evaporation of the supercooled droplets as their water molecules evaporate, and then diffuse toward and condense on the ice crystals. Here is dramatic evidence of the vapor pressure relationships described in Chapter 2 in which at all temperatures below freezing, ice crystals always grow at the expense of liquid water droplets.

A variety of other methods can be used to produce ice nuclei. Laboratory experiments have shown that more than 10 *billion* ice embryos can be formed by the sudden expansion of a cubic centimeter of pressurized air into a supercooled cloud. The sheets of small plastic bubbles used for shipping delicate instruments provide an excellent source, and the sudden bursting of a plastic bubble will generate fantastic numbers of ice embryos. The bubble should be compressed by the thumb and forefinger while held within a supercooled cloud, and then squeezed so as to make a distinct "pop." The high-velocity jet that suddenly emerges from the bubble drops the local temperature by adiabatic expansion to a temperature colder than $-40°C$ ($-40°F$), and large numbers of embryos form by homogeneous nucleation. Quite frequently when ice crystals are formed in this manner they exhibit beautiful colors as they rise toward the top of the chamber. These are interference colors akin to those seen in soap bubbles. The crystals that produce this remarkable effect are extremely thin, featureless, hexagonal plates.

A third method for producing many ice nuclei is to burn a small strip of Eastman Linagraph Paper, which contains silver iodide. A strip 2 mm wide and 3 cm long held by a forceps or similar clamp, ignited above the chamber but then permitted to burn in it, will produce a tremendous number of nuclei. There are many other ways of producing submicroscopic particles of silver iodide but this is one of the easiest and most effective.

A fourth method involves a reaction between polluted air and iodine vapor. A sample of outside air is captured in an ordinary thin plastic garment bag such as the ones used by dry cleaners. This is an extremely effective method for capturing and transporting an air sample; a supply of unused dress-length bags should be obtained. Tie a knot at one end of the plastic cylinder and grasp the other end at either side of its opening; either hold it into the wind to fill it, or swing it in a circular motion if the air is calm. Bringing the edges of the opening together quickly will capture about 50 liters of air.

When the bag of air is injected into the cold chamber, a supercooled cloud generally forms, since the ambient outside air tends to be warmer and much moister than the air at $-20°C$ ($-4°F$). If the cloud is watched carefully and occasionally stirred with a sheet of cardboard, a few crystals may be seen in the light beam. These probably are the "natural" ice nuclei in the air. The concentration of these nuclei ranges from 0.1 to 3 per liter at $-20°C$ in most parts of the world unless the air has been affected by massive dust storms, volcanic eruptions, or other unusual occurrences.

After this observation (which takes from 30–40 seconds) is made, an open bottle of iodine crystals, tincture of iodine, or any other available source of iodine vapor may be introduced and held for a few seconds near the bottom of the supercooled cloud. If the air is then stirred, a very large increase in the ice crystal concentration will be seen. From 1 to 100 or more ice crystals per cubic centimeter will form within 30 seconds; this is 1000–100,000 times more than the concentration of nuclei usually found in ordinary air of the countryside. These new ice-forming nuclei are produced by a reaction between the iodine vapor and submicroscopic particles of lead from automobile exhaust. The residue from the burning of leaded gasoline forms lead iodide, an excellent ice nucleus.

By this extremely simple method it is possible to isolate and capture lead particles with a diameter smaller than 0.02 micron and with a weight on the order of 1×10^{-17} gram. This method of detecting lead in the atmosphere is over 1000 times more sensitive than neutron activation analysis — the most effective method now utilized by scientists studying trace metals in the environment.

Another method for producing many ice nuclei in the cold chamber also employs iodine vapor. A small piece of silver is fastened to a wire connected to one of the terminals of any battery of 6 volts or more. An old silver coin is fastened to the other terminal. If this silver electrode is touched momentarily to the other (thus briefly shorting it), the spark that occurs vaporizes a minute amount of the silver. This metallic vapor condenses to form many submicroscopic particles of silver in the air. The iodine vapor, added previously or afterward to the chamber, reacts to form silver iodide patches on the silver particles, which then serve as excellent nuclei for ice crystal formation.

The same experiment can be carried out with lead electrodes, causing lead iodide particles to be formed.

Note: Only slight traces of iodine vapor should be introduced into the chamber. It is extremely easy for excess vapor to condense on the black velvet lining and to then slowly evaporate, reacting with any lead or silver that may be in the air. Thus, once iodine is used in the chamber, it may be difficult to produce good supercooled clouds.

A seventh method, which comes closest to making ice nuclei that are similar to those found in the natural atmosphere, involves the production of an aerosol of finely divided soil. An ounce or so of powdered soil (clay, sandstone, volcanic ash, and the like) is put into a plastic bottle containing a few pebbles. When the bottle is shaken vigorously a dusty cloud will drift from the mouth of the container. If the plastic bottle is squeezed quickly, a beautiful vortex ring will shoot out of the bottle. If such a ring is projected into a supercooled cloud with a temperature below about $-15\,°C$ ($1\,°F$), a number of ice crystals will appear among the soil particles which slowly diffuse through the clouds away from the vortex ring. These particles apparently act as ice nuclei by becoming "frosted"; after that they grow as misshapen — rarely symmetrical — ice particles.

An eighth seeding method, which involves a still different approach to ice nucleus production, employs a nebulizer. Such a device can be obtained in a drugstore and is normally used for producing a very fine liquid mist for respiratory aid. The nebulizer operates by sending a high-velocity jet of finely divided liquid against a flat surface. This generates many different sizes of liquid droplets. The large droplets fall back into the reservoir to be atomized again, so only very small droplets are expelled into the adjacent air.

If a substance is dissolved in a liquid having a fairly high vapor pressure such as water, alcohol, or other solvent, the dissolved substance can be produced as a very small particle aerosol; when the solvent evaporates, its size can be controlled by the concentration of the dissolved substance.

If a few grams of pentaerythritol (which can be obtained from any chemical supply house) are dissolved in a liter of water, placed in a nebulizer, and aerosolized, the submicroscopic crystals that remain after the water evaporates serve as very effective nuclei for ice crystal formation. These nuclei are extremely interesting; they have unusual electrical properties as nuclei and merit more scientific attention than they have thus far received.

The cold chamber is such a simple device, as are the techniques that can be applied to it, that the reader is urged to try the simple experiments described. There is no better way to understand the complex interrelationships of the factors that control atmospheric processes.

II. The Diffusion Cold Chamber

An interesting, simple, and highly useful cold chamber can be constructed as follows: 4 sheets of heavy window glass $20 \times 45 \times 0.3$ cm thick are fastened together with adhesive tape to make a vertical box and then bonded rigidly with epoxy cement, as shown in Fig. 33. A sheet of Plexiglas $22 \times 22 \times 1$ cm thick serves as the top of the chamber. The bottom is left open. The water supply for the chamber is an absorbent pad $17 \times 17 \times 2$ cm thick made of blotter paper, felt, or sponge rubber. A metal screen fastened with 3 screws holds the pad against the undersurface of the chamber top. A 1-cm diameter hole in this cover provides access to the pad so that water can be added to it as needed. A second 2-cm diameter hole in the center of the top cover provides access to the inside of the chamber. *Note:* Do *not* make the vertical walls of Plexiglas. For some reason not fully understood, a chamber with Plexiglas walls will not operate satisfactorily — the sides must be glass.

Various methods can be used for cooling the lower part of the chamber. If a freezer chest is available (such as the one described on pp. 280–81), the diffusion chamber may be positioned halfway down in it with the open bottom of the glass chamber resting on a wooden frame. This is a convenient arrangement since it provides a completely isolated environment that can be used to hold any type

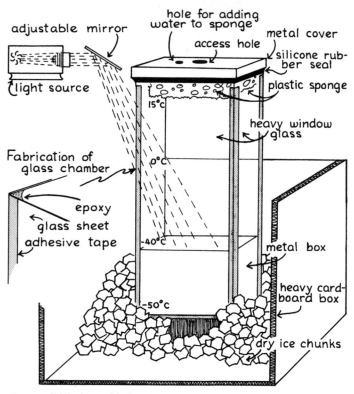

Fig. 33 Diffusion cold chamber.

of air sample desired — the air can be completely free of nuclei or may contain nuclei of a specific type. The bottom of the diffusion chamber, remaining open, provides ready access from below for introducing thermocouples, sampling slides, electrodes, and other equipment needed for experiments. The extremely strong temperature inversion (described below) makes the air in the chamber very stable.

The diffusion chamber can also be made as a separate entity. The cold source can be provided in several ways, but the one recommended uses dry ice (solid carbon dioxide) as the coolant. Fragments of dry ice, which has a temperature of $-78.5\,°C$ ($-118\,°F$), can easily provide a temperature in the lower part of the chamber colder than $-40\,°C$.

If a chamber is constructed as described and illustrated here, its equilibrium temperature is about 20°C (68°F) at the top, while the bottom is −50°C (−58°F). The temperature range is similar to that at the middle latitudes of the earth between its surface and the base of the stratosphere. Thus it is possible to simulate cirrostratus clouds at the bottom formed by homogeneous nucleation; altostratus clouds in the midportion, having temperatures ranging from 0°C to −20°C where heterogeneous nucleation must occur if ice crystals are to form in the normally supercooled clouds; and warm stratus clouds or fog between the 0°C level and the top of the chamber.

To provide a self-contained unit, it is necessary to build a base of heat-conducting material such as brass or aluminum. The ideal arrangement is to place a sheet of brass or aluminum 30 × 30 × 1 cm thick at the bottom of a square, heavy corrugated cardboard or wooden base 25 cm high with inside dimensions of 35 × 35 cm. A solid cylinder of brass or aluminum 15 cm in diameter and 10 cm high rests on the metal plate and becomes the heat sink that cools the lower part of the chamber.

The bottom of the diffusion chamber is made of a sheet of brass 22 × 22 × 1 cm thick. Soldered to the top of this is a box made of sheet brass 0.2 cm thick whose outside dimensions are 18 × 18 × 20 cm high. This box is painted a flat black after being soldered to the base plate; it conducts heat from the lower part of the chamber to the heat sink.

It is quite possible to reduce the size of the diffusion cold chamber to about half that of the one described, with all dimensions reduced in proportion, but the larger size is preferable.

One of the most interesting experiments in the diffusion chamber, after it has come to temperature equilibrium, is related to dry ice seeding. A few tiny fragments of dry ice are dropped through the cloud in the chamber. Condensation trails form instantly, and then within a few seconds a vortex ring develops at the 0°C level of each trail. The upper trails and the vortex rings formed by their descent consist of water droplets, while the trails below 0°C are made up of twinkling ice crystals that rapidly grow and develop dark sheaths around them as they pull moisture from the nearby supercooled droplets.

Vortex rings containing ice-forming nuclei, hygroscopic particles, and other chemicals can be introduced through the access hole in the top of the chamber.

After the chamber has been in operation for several hours, its cloud will decrease in density and visibility until it finally disappears. This happens because the supply of nuclei in the air of the chamber is exhausted. The air then becomes highly supersaturated so that if a few nuclei are injected by the vortex-ring technique into the center of the chamber, they will grow very rapidly and fall with increasing downward velocity. Other important and equally interesting experiments can be performed with this chamber.

III. The Preparation of Snow Crystal Replicas

It is easy to preserve the shape and beauty of snow crystals. One gram of polyvinyl Formvar resin* is dissolved in 100 milliliters of ethylene dichloride to make about a 1% solution of the plastic resin. A glass slide is wetted with this liquid after slide and solution have been cooled below 0°C. The wet slide is then held in the falling snow. After a sufficient number of crystals have fallen on the slide, it is placed in an area protected from the snow but colder than 0°C until the solvent has evaporated. This may take 3–5 minutes or longer. (The Formvar solution has the interesting property of creeping up over a snow crystal that may rise considerably higher than the depth of the liquid on the slide.) When the solvent has evaporated, the crystal is encased in a thin plastic shell. Once the slide is dry, it may be warmed; the water then passes through the plastic shell leaving behind an exact replica of the surface features of the original snow crystal. If the crystal has a particularly fine structure, it is best to let the ice evaporate rather than melt; melting sometimes produces surface tension that may cause the collapse of delicate features in the crystal.

Another effective method for making replicas is to use the aerosol type of clear plastic spray available in most paint and hardware stores. The aerosol spray is most useful for replicating windowpane frost and similar ice structures. The spray can must be positioned so that the surface to be replicated is only slightly moistened by the spray deposit — the solvent used in some of these solutions tends to etch or dissolve the ice if too much liquid is present.

It is also possible to make good snow and ice replicas by the plastic spray method. The best procedure is to precoat the glass slide with the plastic film at room temperature. The slide can then be placed either in a cold chamber so that seeded crystals fall onto it, or held out in a natural snowstorm. After enough crystals have landed on it, the slide is sprayed again until the surface moistens. After the spraying stops, the solvent quickly evaporates. The slide may then be warmed for inspection.

A third method for preparing replicas of snow or other frozen particles in a winter storm permits the selection of particularly beautiful or otherwise interesting samples of the precipitation. (It is unusual for all crystals of a snowstorm to be unbroken or otherwise symmetrical.) A 1% solution of Formvar is cooled to $-5°C$ (23°F) and placed on a table or shelf in a cold sheltered place. A 20×30 cm (8×12 in.) piece of black velvet mounted on a sheet of cardboard is held in the falling snow and then examined for crystals or particles of interest. A toothpick or similarly shaped wooden rod is dipped into the Formvar solution and withdrawn quickly so that a drop of liquid drains to the end of the rod. The

*Resin 15-95E, Monsanto Chemical Corp., Springfield, Mass.

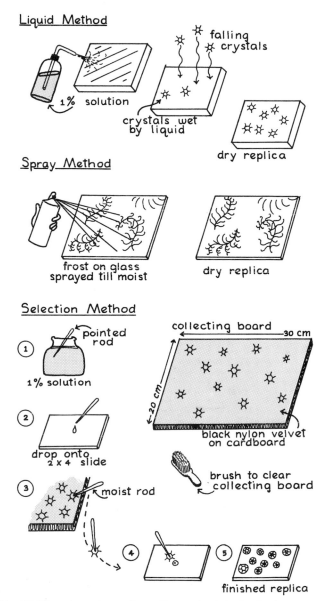

Fig. 34 Ways to make replicas of snow crystals, frost, etc.

drop is laid on the cold glass slide. The rod, still wet, is touched lightly to the crystal selected for replication. The crystal adheres to the pointed rod and is then immediately transferred to the drop on the slide, leaving the rod and sinking into the liquid. Any number of crystals can be selected and transferred to the glass slide in this manner.

This method is particularly convenient if one wishes to project replicas in a 5 × 5 cm or other type of slide projector. A second glass slide or plastic cover may be placed over the replicas to protect them from dust. Replicas formed by this method are tough and will resist damage.

Cross sections of hailstones, icicles, hydromites, and other solid ice structures may also be replicated. The ice is cut with a band saw while very cold, and its surface is smoothed and then held in a cold region until some evaporation occurs. The molecules of ice evaporate from the surface, leaving behind a beautiful pattern of etch pits. A 1% Formvar solution is flowed over this etched surface, allowed to dry, and then peeled from the ice and mounted on a glass slide.

For those not fortunate enough to live in snow country, replicas can be made by making frost in the freezer compartment of a refrigerator or in a mixing cold chamber (see pp. 280–82).

IV. Bubbles for Atmospheric Studies

The ordinary soap bubble is a valuable tool for measuring certain features of the atmosphere. The best bubble solution for most of our purposes is the cheapest, containing a little surface-active material like soap and a great deal of water; it is sold in variety stores.

An obvious use of the soap bubble is observation of local airflow patterns. An hour spent every day for a week in observing the motions of bubbles formed outdoors will show the complex flow patterns of the local air. Turbulence, laminar flow, eddy diffusion, wake effects, and convection on a small scale are made beautifully visible. With a little experience, bubbles of controlled size and number can be produced using various-sized bubble rings and airflows.

A most interesting phenomenon can be observed when large bubbles are made in temperatures colder than $-10°C$ ($14°F$). Shortly after a large bubble starts floating in the cold air, one or more ice crystals are likely to start growing on its surface; this is caused by the presence of ice nuclei or tiny ice crystals in the air. The crystals in the bubble film grow rapidly until the bubble either breaks or becomes completely frozen. Quite often, when a number of crystals form and the bubble breaks, the crystals fall separately, and by counting them it is possible to ascertain roughly the number of ice nuclei in a given volume of air. Large differences are often encountered.

It is easy to see how supercooled bubbles or bubble films can be used in a cold chamber. If a bubble is formed, captured on a ring, and then lowered into a cold chamber, it can be used to detect the presence of even a very few ice crystals in the chamber. Thus the number of natural ice nuclei in the atmosphere can be measured easily by introducing a sample of atmospheric air into the chamber, forming a supercooled cloud, and then observing the surface of the bubble for ice crystal formation.

At least 3 basic forms of crystals will be observed growing in a soap film. One is in the form of a hexagonal dendritic crystal — this has the normally 6-sided crystallizing pattern of ice. Another appears in the shape of a cross, indicating the columnar or needle form with only 2 of the major crystalline axes. The third is an asymmetric frostlike crystal. This is an irregularly shaped crystal of the type that generally forms on natural soil particles which act as ice nuclei.

A more convenient application of this technique is to form a flat soap film by dipping a ring, square, or rectangle made of thin wire into a shallow tray containing the soap bubble solution. When the wire form is withdrawn, it will bear a flat film that, when put in the cold chamber, quickly supercools and then serves as an ice nucleus or crystal indicator.

Fig. 35 Ways to produce ice crystals on supercooled bubble surfaces, flat films, and thin layers of water.

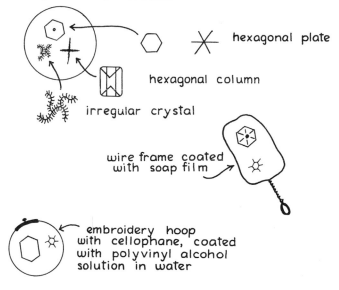

hexagonal plate

hexagonal column

irregular crystal

wire frame coated with soap film

embroidery hoop with cellophane, coated with polyvinyl alcohol solution in water

An interesting demonstration of the presence of the inversion above a cold chamber can be made with a free-floating bubble. If a bubble is formed and allowed to drift down toward the open chamber, it will suddenly stop falling as it contacts the inversion. It will bounce a few times at the inversion interface and then slowly sink into the cold air and fall to the bottom. If the bubble is properly illuminated, the inside of it will be seen to contain a fog. Generally the bubble, especially if it is small, will freeze at the bottom of the chamber, sometimes holding its form for a long time. After the ice has evaporated, the bubble may be held together by the soap skeleton, which remains intact. This was formed when the freezing front of the ice crystal concentrated the soap of the bubble solution in a lacelike pattern.

This observation suggests another highly useful technique. If a water-soluble substance is dissolved in water, spread as a thin coating on a sheet of black cardboard, a glass plate, or a piece of cellophane stretched on an embroidery hoop, and then lowered into a cold chamber or placed in cold air outdoors, the water in the solution will supercool and when contacted by one or more ice nuclei, will begin to freeze in the same patterns as those observed in the soap bubble. As the crystals grow, the ice drives the soluble chemicals ahead of it, concentrating them until they precipitate out as a visible residue. Some of this may be engulfed by the dendritic habit of the crystals, although the bulk is finally concentrated at the boundaries between the crystals as they merge.

If this frozen film is left in the cold chamber for a day or so, the ice will evaporate completely, leaving behind a skeleton framework of the chemical which outlines the former crystalline ice in a beautiful pattern.

In addition to being quite beautiful, these structures provide an excellent permanent record of the number of crystals or nuclei and their pattern at the time that the replica was made.

One of the best materials for coating the surface in this experiment is water-soluble polyvinyl alcohol.* Other substances may be used, ranging from salts to sugar and gelatin. Some will modify the crystalline habit of the ice in a major way; others will not provide very good contrast. There is plenty of room for new discoveries in this process.

An extremely useful and exceedingly tough bubble can be made from equal quantities of glycerine, white corn syrup, and concentrated liquid detergent. The latter substance is an uncertainty because of the frequent changes in formulae, so experimentation is required. The detergent is used for its film-forming properties, the glycerine for longevity, and the syrup for toughness. If you are successful in forming a stable film with such a mixture, you will then be able to make huge bubbles. They can be made nearly a

*Elvanol 7584, DuPont Chemical Corp., Wilmington, Delaware

meter in diameter by placing the viscous solution in a large flat tray and using a hoop made from a wire coat hanger. The hoop is lowered horizontally into the solution, and a flat film is brought out on the hoop and then moved through the air, holding the flat film vertically. One can often pull a bubble cylinder of 2 meters or longer, with a diameter equivalent to or larger than the coat hanger hoop. Such films are extremely stable and are ideal for airflow studies. They are no good at all for the ice crystal detection studies, however, because the solution contains very little water. When they break the falling residue appears similar to a collapsed parachute. Obviously such bubbles should not be used inside the home!

One final technique related to bubble experiments: Some of the useful features of bubbles can be simulated with a thin polyethylene or other type of plastic bag. If 25 cubic centimeters of water mixed with about 5 drops of detergent is placed inside a 10-liter bag, it will wet the inside surface as an air sample is captured. The bag can then be taken to a cold chamber and lowered into it, and a supercooled cloud will form as the air cools. Any ice nuclei present in the sample air can be detected either in the beam of light or by the way the thin film of water freezes, coating the internal surface of the bag.

If another bag is prepared in a similar manner, but in addition a few iodine crystals are placed in it, a greatly increased number of ice nuclei will be seen as explained on p. 284.

The value of this technique lies in the freedom it gives the observer to prepare or capture an air sample under specialized conditions even a considerable distance away from the cold chamber, and then to study reactions in the isolated air sample under precise control.

V. The Detection of Submicroscopic Particles

A very interesting and effective method for detecting the presence of invisible particles in a sample of air applies the principle of the Wilson Cloud Chamber. This, in its simplest form, introduces an air sample into a container such as a glass bottle to which a small amount of water has been added. The air sample is put under pressure and then the pressure is suddenly released, forming a cloud in the bottle. The sudden release of pressure causes the moist air to cool and for a short period to become supersaturated with water vapor; the degree of supersaturation depends on the amount of pressure applied. The cloud that forms is produced as the excess water in the supersaturated air is deposited on the suspended particles. If no particles were present, this water would be deposited on the walls of the bottle. The cloud that forms has a concentration of water droplets directly related to the size of the original nuclei and the degree of supersaturation produced by the expanding air. If the supersaturation in the bottle is about 1%, only the larger cloud condensation nuclei form the visible cloud. If a high pressure — about 50 cm (20 in.) of mercury — is used, the supersaturation will be 300% and practically all of the submicroscopic particles down to an equivalent diameter of 0.005 micron will be forced to accept water. A dense cloud will probably be seen.

In the polluted air of a city, the cloud will be extremely dense at the high supersaturation level and even at low supersaturation. In fact, a low supersaturation cloud in polluted air will appear to have about the same scattering power in a beam of light as the cloud that forms in a country air sample subjected to high supersaturation.

The cloud chamber will work equally well if the humidified air sample expands into a vacuum rather than dropping from excess to normal pressure. In both instances the cloud forms by adiabatic expansion.

If a cloud chamber is assembled so that a vacuum of 2.5–50 cm (1–20 in.) of mercury can be developed and moist sample air then permitted to rush suddenly into the vacuum, a cloud will form. If the container is fitted with a vacuum gauge, a vacuum of 2.5 cm of mercury will produce about a 1% supersaturation of water; 50 cm will produce 300%.

A very inexpensive and highly satisfactory demonstration or observation chamber can be made using a glass aquarium with a volume of 2½ liters, 2 sides flattened and 2 rounded, as shown in Fig. 36. Such a container has a large hole at the top that can be closed using a disk of plywood sealed with plasticene. Three holes cut into the wooden disk provide a means for introducing the air sample, attaching the vacuum pump, and holding a vacuum gauge. Otherwise this hole, which is larger than the others and is plugged with a large cork or rubber stopper, may be used for cleaning.

A strong pinch clamp on the air sample tube may be connected to an 8–10-liter plastic sample bag that contains the air to be observed. A bicycle pump fitted with a simple vacuum valve can be used to produce the vacuum. A tire valve in reverse position will probably be adequate. The leather disk in the bicycle pump must be reversed in order to pull a vacuum.

One of the flat sides of the aquarium should be painted black on the outside to improve contrast for observing the cloud. The cloud that forms through expansion should be illuminated with a strong parallel beam of light, such as a slide projector. The container should have about 25 cubic centimeters of water in it mixed with a small amount of detergent, soap, or other surface-active material so that when it is sloshed around in the container, the inside surfaces remain wet.

While the unit described is useful primarily for demonstration and qualitative observations, with a little experience and ingenuity

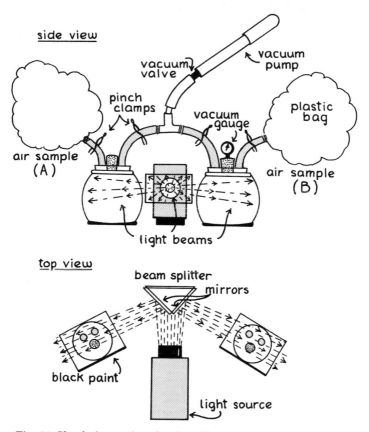

Fig. 36 Cloud observation chambers (for comparing samples of air).

plus a basic understanding of the principle involved, it is possible to construct an instrument that will provide comparative, semi-quantitative, or even quantitative measurements of small particles in the atmosphere.

For comparative measurements, 2 identical aquarium-type observation units should be used. These are evacuated with a common vacuum pump fitted with a T-connection and can be used to compare 2 air samples. Pinch clamps located as shown in Fig. 36 permit separate manipulation of the 2 units. A common light source, whose beam falls symmetrically on mirrors, can be used to illuminate both chambers.

For more quantitative measurements, an adjustable-intensity

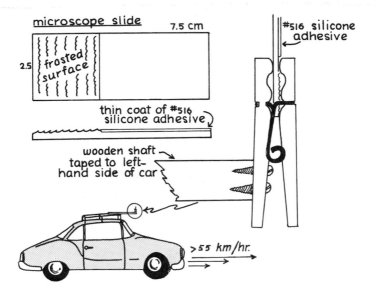

Fig. 37 How to catch airborne particles such as pollen grains, soil particles, and fibers.

light source should be used connected to a 2.5-cm-diameter cloud chamber having a path length of 50 cm or more. A photocell is located at the end opposite the parallel (or slightly converging) light source. The photocell is connected to a microammeter and the current produced by the light cell is adjusted to register a full-scale displacement when the chamber is free of cloud. When a cloud forms, the light scattered by the cloud produces a reduction in the light received by the photocell. This can be calibrated in terms of the number of nuclei per cubic centimeter using a Small Particle Detector (see p. 217). These fine particle detectors are now used in air quality surveys.

VI. Catching Large Aerosol Particles in the Atmosphere

There are many different kinds and sizes of airborne particles in the atmosphere. The smallest ones are the most numerous; many are invisible. They are formed by the condensation of gaseous material from both manmade and natural sources. Such particles rarely exceed 1 micron in diameter.

The larger particles floating in the air are relatively few but of great variety. They include such things as tiny fragments of rock, spores, pollen grains and other organic substances, agglomerations of smoke particles, fragments of burned material of all kinds, spider webs, and small insects.

There are various ways of obtaining representative samples of these large particles. Since there are not many per unit volume, quite a bit of air must be sampled to find them. One of the easiest and most successful methods is to coat the surface of a glass microscope slide with a sticky substance, hold it in a clamp fastened to the end of a stick, and fasten the stick to the top of a car so that it extends beyond the slipstream of the automobile (see Fig. 37). For each kilometer of travel about 1.5 cubic meters of air is sampled. A good coating of particles can be obtained in a distance of about 17 km (10 mi.).

The sampling slides can be prepared from 2.5 × 7.5 cm (1 × 3 in.) microscope slides (the type that has a third of the surface frosted is preferable). The slides are given a very thin coating of vaseline. To get a uniform distribution of particles across the slide it is necessary to drive over 60 km/hr. (40 mph); at slower speeds there is a tendency for the sample particles to be at the edges of the slide, with very few in the center.

Photomicrographs of collected particles are easily prepared and will lead to many surprises. There is a great need for more information about the types of large particles to be found in the air of various parts of the globe.

VII. The Photomicrography of Snow Crystals

Shortly after the turn of the century, Mr. Wilson Bentley, a farmer living in the snow country near Jericho, Vermont, produced thousands of beautiful photomicrographs of snow crystals. Mr. Bentley used simple photographic equipment in his studies. A modern version of his technique utilizes inexpensive equipment not available to him and further simplifies the procedure he followed.

The basic unit for snow crystal photography is a low-power microscope. These are now available at a modest cost from several supply houses (such as Edmund Scientific Corp., 300 Edscorp Bldg., Barrington, N.J., and Monroe Microscope Co., P.O. Box 656, Rochester, N.Y.). At the Rochester company and other similar firms, excellent reconditioned microscopes are sold.

The primary requirements for a microscope are that it be easy to focus and that low-power objective lenses be available with a "universal" mount. A total magnification of 10X or more is essential for natural snow crystals, although it is desirable to have higher magnification lenses for preparing pictures of crystals formed in the cold chamber. A magnification of about 200–500X is needed for the ice crystals that form in the laboratory.

Whatever microscope lenses are obtained, they should be of good quality and reasonably free of spherical and chromatic aberration. This means that an object viewed under the microscope should be free of distortion and in sharp focus toward the edges of the view, and that the object should not display spurious colors near its edges. If funds permit, the microscope should be of "standard" size.

One factor essential for successful snow-crystal photography is proper illumination. Most light sources are too hot and cause the crystals to evaporate or melt. A very satisfactory source for both viewing and photography is the type of bulb used in the so-called "penlite" miniature flashlight of 1½ or 3 volts. This bulb has a

Fig. 38 Photomicrography of snow crystals.

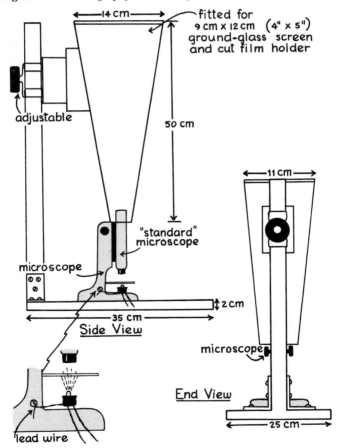

small spherical projection lens as part of its structure that produces a nearly parallel (just slightly divergent) beam of light. This light source does not produce enough heat to damage even a very delicate ice crystal and yet provides enough illumination as transmitted light for exposures on the order of 1–10 seconds. The light bulb should be mounted under the microscope stage on an adjustable suspension so that it can be moved about until the best possible contrast and illumination of the crystal is seen on the ground-glass viewing screen.

A tapered, light-tight box serves as an extension chamber. It can be made of 1-cm thick plywood with a hole in the bottom end that fits freely over the tube of the microscope. A black velvet or felt lining makes this connection light tight. The inside of the tapered extension box should be painted a dull black or lined with black velvet cloth.

The plywood box is made to the approximate dimensions shown in Fig. 38 and mounted on an adjustable wooden arm that in turn is fastened rigidly to a sheet of 2 cm marine plywood cut to 25×35 cm (10×14 in.) which serves as a base. The top of the extension box should be fitted with a light-tight adapter that will accommodate (1) a ground-glass screen, (2) 9×12 cm (about 4×5 in.) cut film, 9×12 cm film pack, or a 9×12 cm Polaroid-Land film holder.

It is very convenient to have the ground-glass screen arranged so that it can be covered with a 25–30 cm (10–12 in.) high light-shield to assist in focusing the image under the microscope. A 2X hand lens is often of considerable assistance in making a final sharp adjustment of the image at the focal plane (where the photomicrograph will be exposed).

For sampling natural snowstorms it is preferable to have the equipment in a sheltered place outdoors. The best results are obtained at night, especially when the snow is not falling heavily. The view of the crystals for sharp focusing and all other operations favors making pictures after dark.

A black velvet cloth mounted on heavy corrugated cardboard serves as a good sampling board. A whisk broom should be available to clean off the velvet before obtaining a new sample.

The easiest way to select a crystal is to place a cold microscope slide on top of the specimen resting on the black velvet surface. Pressing the slide lightly against the crystal will generally make it adhere to the glass surface, and it can then be placed uppermost on the microscope stage for observation and photography. If the air is extremely dry and there is a tendency for the crystal to evaporate, a cylinder of cardboard 2 cm in diameter, 0.5 cm thick, and 0.5 cm high should be soaked in water and frozen and then placed symmetrically around the crystal. This provides a saturated microenvironment that will protect the crystal for a long time.

The same photomicrographic device can be used in a cold chamber by placing the plywood base on the bottom of the freezer chest. The microscope should be entirely immersed in the cold air; otherwise, condensation will occur on any metal projecting above the top of the inversion.

For photographing the crystals that form in the cold chamber, higher magnification is needed. A 25–40X objective and a 20X ocular lens is generally good for viewing. It is desirable to have a variety of objective lenses and eyepieces. They do not need to be of professional quality, but good secondhand optics of high quality are available at a modest cost from microscope supply houses.

For film, Kodak Tri-X Panchromatic ASA 400 is preferable, either as cut film or film pack. Polaroid P/N #55 ASA 50 and #52 ASA 400 also give excellent photomicrographs. Polaroid #57 ASA 3000 may be useful if very high magnification is needed. Neither of the last two provides negatives for making additional prints.

For the sake of economy, slides smaller than the 9 cm × 12 cm dimensions recommended may be used. This may be a particularly good idea for beginners who are just developing their technique. Adapters for holding 5.6 cm × 8.2 cm (2¼ × 3¼ in.) cut film are available that will fit into the 9 × 12 cm socket of the equipment constructed. With experience the microscopist will find that for most purposes the larger film is preferable. It will provide excellent contact prints and enlargements are superb!

For those who are not familiar with the use of film, film holders, and related film and print processing, there are many excellent books and pamphlets available at camera stores. Personnel at such stores can be very helpful. The basic principles of photography are very simple, easily applied, and can be learned quickly.

VIII. The Formation of Clouds in a Water Cloud Box

A series of interesting and instructive demonstrations of the features of cloud development can be carried out in an aquarium filled with tap water. A small aquarium about 30 cm wide × 45 cm long × 30 cm deep, with a capacity of about 36 liters (9 gals.), is the size recommended.

Prior to filling the tank, a plastic tube 2–3 mm in diameter is fastened vertically along one corner of the aquarium and then led to the center of the bottom surface. There it is fastened to a glass or plastic orifice having an inside diameter of 1–2 mm. It is very easy to make orifices, capillaries, bent tubes, and similar objects ordinarily limited to the skill of a glass blower by using polyethylene tubing rather than glass. Held in the low flame of a propane torch — obtainable in a hardware store — the translucent tubing is heated until it becomes transparent. It can then be bent, drawn, or blown like glass, but at a much lower temperature.

The tank is initially filled with 27 liters of cold tap water. The remainder of the tank is filled with hot water (about 9 liters at 50°C), which is floated on top of the cold. This is easily accomplished by laying a 28-cm diameter, 1-cm thick disk of marine plywood on top of the cold water. When the hot water is slowly poured on the middle of the floating disk, it dribbles onto the water below and little mixing occurs since the hot water is more buoyant than the cold.

After the tank is filled with water, a simulated cloud can be formed as follows: 200–500 milligrams of crystalline silver nitrate is dissolved in a liter of distilled or de-ionized water. A small amount (25 cc) of methyl alcohol is added to the water to give it positive buoyancy. This water is then placed in a plastic bottle with a hole in the bottom, into which is inserted the upper end of the flexible tube that is fastened into a corner of the aquarium. (Water-tight connections of this sort can be made by cutting a round hole in the plastic bottle 1 mm smaller than the outside diameter of the tube to be inserted.)

Raising or lowering the bottle containing the dilute silver nitrate solution controls to a limited degree the flow through the tube. However, it is handiest to use a squeeze clamp with a fine adjustment at least during the initial phases of the experiment. The adjustment of this clamp in combination with the height of the silver nitrate solution provides quite a bit of flexibility.

A stream of distilled water containing the silver nitrate is now

Fig. 39 Water cloud box.

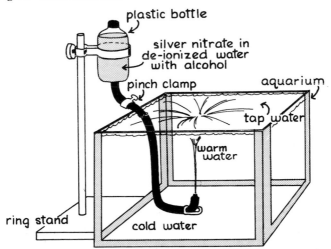

released into the aquarium from the orifice at the bottom of the tank as a slowly rising stream. The silver nitrate reacts with the chloride ions in the tap water to form a visible precipitate of silver chloride, which will appear bluish (if the silver nitrate solution is very dilute) to pure white (if the reacting materials are more concentrated). If a test indicates that the chloride content of the tap water is low, this can easily be rectified by adding some common table salt when the tank is being filled. One cubic centimeter of tap water combined with an equal amount of silver chloride solution will produce a white precipitate if the solutions are of proper concentration.

The most satisfactory reaction produces a slightly transparent but noticeably white cloud. A word of caution — silver nitrate stains the hands if used in concentrated form.

When silver nitrate solution flows into the tank, it may rise upward for a short distance and then descend to the bottom. If this occurs, some additional methyl alcohol should be added. When the vertical flow is very slow, a remarkable series of mushroomlike structures — in fact a series of horizontal vortex rings — will form as the plume works its way to the top. The way they diffuse into the surrounding water provides fascinating material for discussion and thought as to how such formations relate to the atmosphere.

Interestingly, the various patterns that can be formed in the water system can all be produced in the diffusion cloud chamber as well.

If the flow of silver nitrate solution exceeds a certain velocity, a shift from laminar to turbulent flow will be observed. When it reaches the turbulent stage, a structure that looks very much like a cumulus cloud will appear at the top of the chamber.

Under most conditions a very slow rotational component will appear and will eventually (after several hours of slow flow) produce a 3-dimensional structure that looks very much like a hurricane. Even precipitation elements may appear after a time, or they may be simulated with very fine grains of sand.

Once the experimenter gets a feel for the several rather simple components of the cloud box — liquid stability or instability, the velocity effects of the silver nitrate stream, the relative concentration of silver and chloride ions, and the effect of light on the color of the silver chloride colloidal suspension — a fascinating series of provocative and highly instructive experiments can be conducted.

In addition to gaining a better understanding of atmospheric processes from these observations, there is a great possibility of making new discoveries with them.

IX. Vortices of Air, Fire, and Water

Vortices in the atmosphere range in size from snow or dust devils and fire whirls to the waterspout, tornado, and fire storm, to the massive organized storm systems epitomized by the hurricane and typhoon over the oceans and continent-wide low pressure storms over the land. For experimental purposes we will limit ourselves to modeling the tornado, the whirlpool, and the fire whirl (this latter structure is somewhat akin to the dust devil). They all have features in common.

The most impressive demonstration simulates a tornado and requires a circular metal container with a diameter about ⅓-½ of its height. The smallest practical size is a 4-liter (1 gal.) can; the largest is an oil drum with a capacity of 220 liters (55 gals.). The other items needed are an old-fashioned vacuum cleaner that sucks air in a swirling motion, enough hot water to cover the bottom of the tornado chamber 2 cm deep, and a projection-type floodlight for illuminating the inside of the chamber. Fig. 40 shows a unit made from an oil drum.

A rectangular hole is cut into the side of the container. This should be centrally located; its height should be about half the height of the can, and its width about half the diameter of the can. Two round holes are also made in the top, one in the center for the suction port of the vacuum cleaner, the other for mounting the projection lamp. For a 10- or 15-minute operation, the water in the bottom of the container can be heated to the boiling point before it is poured in.

As the velocity of the motor increases, an air vortex becomes visible. When properly formed it looks very much like a real tornado or waterspout, and a miniature column of water will actually develop where the vortex contacts the surface of the hot water.

If a chunk of dry ice is placed in the hot water, the vortex tube becomes quite spectacular. A similar effect occurs when a large number of condensation nuclei are provided, which can be done by striking a kitchen match.

A number of interesting experiments can be carried out with the vortex generator once its basic physical relationships are understood. For example, if the unit is placed in a large container fitted so that the air exhausted by the vacuum cleaner is passed again through the vortex, any airborne particles are subjected to intense "scrubbing" forces. Thus the scavenging action of a moist air vortex can be studied.

To form a beautiful whirlpool, a plastic cylinder or box with the dimensions shown in Fig. 40 provides a good starting point for experimentation. If a hole is cut in the bottom of the chamber large enough to accommodate a 2.5-cm (1 in.) cork or rubber stopper, various sizes of drain orifices can be employed to control the dimensions of the vortex. A tangential jet of water at the top of the tank drives the vortex. If an adequate supply of water is available, a tube providing water to the jet and a drain below the chamber to dispose of the water are all that is needed to carry out the demonstration. If the water supply is limited, a small water pump con-

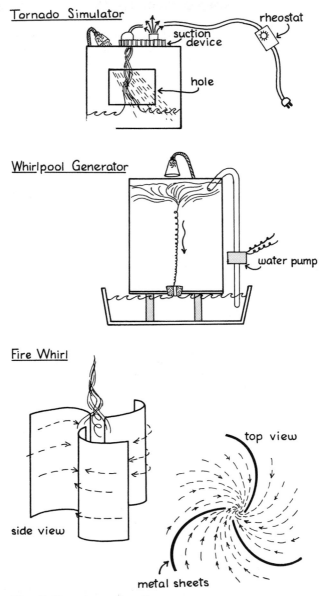

Fig. 40 Three types of vortices.

necting the jet to the drain of the pan below the vortex chamber provides a closed water cycle. It is amazing how many different types of whirlpool vortices can be generated by the clever experimenter.

The third vortex demonstration is the fire whirl. In its simplest form the fire whirl may be demonstrated by using the tornado chamber. Instead of water, a can of Sterno or some other alcohol-based fire starter may be used. A large candle, a stick of burning incense, or a kerosene-fed cotton wick may be used. If large particle-producing fires are created, smoke will eventually fill the room.

Another type of fire vortex uses a fire of burning paper, cardboard, wood, pine cones, or other fuel in an area where the air enters in a tangential direction as shown in Fig. 40. Such a fire whirl closely resembles in miniature the gigantic fire whirls that often occur in a forest fire, and are dreaded by the forest-fire control experts, in which the air enters the fire at a tangent. Such fire vortices sometimes become large enough to carry burning branches into the sky and scatter them over mile-wide areas. These can cause the "blowup" — the worst kind of forest fire.

X. Simple Devices for Measuring Wind Velocity

Wind is an interesting and important phenomenon in the atmosphere. The measurement of its velocity at ground level is a good way of acquiring a better understanding of the global circulation, especially if the local winds at ground level are observed in relation to newspaper or television weather maps that show the position of high and low pressure cells, the path of jet streams, and winds generated by local storms.

A variety of devices are available for measuring air velocity and direction. They range from a device in which the passing wind causes a pith ball to rise in a tube that is calibrated to show wind velocity (Dwyer Windmeter, F. W. Dwyer Mfg. Co., Michigan City, Ind.) to highly sophisticated instrumented anemometers, aerovanes, and vibrating reeds that provide electrical impulses to actuate dials, recorders, or digital tapes.

One of the simplest of all devices for measuring wind velocities up to about 30 meters per second (67 mph) is a bivane made of a 7.5 × 12.5-cm (3 × 5 in.) library file card, a paper clip, a wooden toothpick, a piece of string, a bit of glue, a piece of lead wire, a sheet of cardboard, and a plumb bob (see Fig. 41).

The file card is bent to form an included angle of about 60°. This angle is preserved with a toothpick glued to the 2 sides; it is also possible to devise a spacer that can be removed or collapsed.

The paper clip is partly straightened with its straight part glued along the bend in the cardboard, and a thread of nylon cord or monofilament is fastened to the center of the bend. At the lower part of the bivane the end of the wire clip is bent at right angles

Wind Measurement

side view

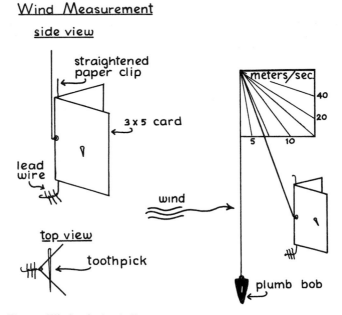

Fig. 41 Wind velocity indicator.

and the lead wire wrapped on it. This is used to counterbalance the weight of the wings of the cardboard and toothpick. When properly weighted, the supporting string rides parallel with the front face of the bivane in calm air.

When the wind blows, the bivane rides the wind without fluttering if it has been correctly made and balanced. Its string should then be mounted at the top and at one side of a thin sheet of cardboard or other light and flat material on which are marked angles from 0° to 90° at 2° intervals as shown in Fig. 41. The plumb bob hangs along the vertical side of this sheet to keep it properly oriented. When the wind causes the bivane to sail, the angle of its supporting filament is related to the wind velocity. The unit can be calibrated by having someone hold the assembly out of a car window beyond the slipstream and note the angles of the filament that correspond with the velocities indicated by the speedometer. These angular displacements can subsequently be converted to meters per second.

With very low wind velocities, it is possible to determine the drift of the air by throwing a handful of dust or snow into it, or by holding up a wet finger. The wet finger becomes coldest (by evaporative cooling) in the direction from which the air is moving. A simple floating dial compass can provide a very useful indication of the wind direction. *Note:* In recording airflow in atmospheric studies, the direction given is always the one from which the air is coming. Thus, W–5 indicates the air is coming from the west at 5 m/sec. (10 mph).

A spider web fastened at one end to a stick and at its other end to a seed parachute, such as dandelion, salsify, or milkweed, is an extremely sensitive air velocity indicator.

XI. The Measurement of Temperature and Humidity

The measurement of temperature and relative humidity is most simply done with a sling psychrometer. More expensive and elaborate modifications of the wet and dry bulb principle are available but do not give any better data than can be obtained with carefully used wet and dry bulb thermometers.

It is quite feasible to mount 2 straight thermometers side by side as shown in Fig. 42. The temperature bulbs should extend into space beyond the holder, which can be made of wood, metal, or even cardboard.

The wet bulb thermometer extends about 3 cm beyond the end of the dry bulb, which in turn is 2 cm beyond the edge of the holder. It is extremely important to have the thermometers anchored firmly to their holder, since the centripetal force is quite high when they are swung.

A piece of thin cotton gauze is tied with cotton thread beyond and above the bulb of the thermometer to be used for measuring the "wet" temperature.

A strong nylon string or light chain 15–30 cm long, fastened to the end of the thermometer holder opposite the bulbs, is in turn fastened to a small handle as shown in Fig. 42.

To obtain a reading of air temperature and relative humidity, the cotton jacket of the wet bulb thermometer is thoroughly moistened with distilled or de-ionized water. Tap or stream water or even saliva may be used in an emergency, although it evaporates and leaves a residue. The assembly is then swung rapidly for a minute or two. For this operation it is desirable, if possible, to be in the shade of a tree or building, although care must be exercised that the air being measured is not modified by some noticeable local effect. At times it is possible to use one's own shadow. The place selected for making the measurement should be free of obstacles so that the thermometer will not hit anything.

After the psychrometer is swung for 30 seconds or so, the swinging motion should be stopped as quickly as possible and the wet bulb read immediately, then the dry. It is best to write down the values in a pocket notebook — though the palm of the hand is a convenient surface when nothing else is available. After noting the initial readings, the instrument is swung again and a second reading made. This sequence is continued until the values do not change. With both wet and dry bulb values ascertained, it is a simple matter to calculate the relative humidity. Slide rules and nomographs are available or may be constructed for this purpose. (See Appendix 15.)

A crude but simple method for ascertaining the dew point involves a metal container filled with water and a small block of ice. The container is filled with the ice and water and stirred with a thermometer. The outer surface of the container should be felt each time the thermometer is read. At a specific temperature, the surface of the container will become wet. This is the dew point temperature of the surrounding air. If the dew point is particularly low due to very low relative humidity, some rock salt can be added to the crushed ice or snow in the container. A temperature as low as $-16°C$ (5°F) can be achieved in this manner.

Fig. 42 Sling psychrometer.

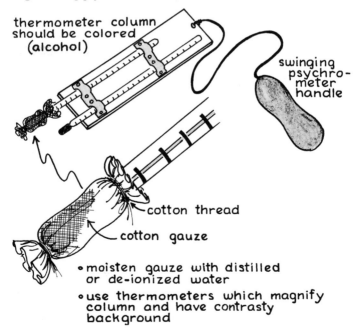

thermometer column
should be colored
(alcohol)

swinging
psychro-
meter
handle

cotton thread

cotton gauze

• moisten gauze with distilled
  or de-ionized water
• use thermometers which magnify
  column and have contrasty
  background

XII. The Sampling of Rain

Everyone has seen raindrop splash marks on the pavement, rocks, paper, windshields, windows, and similar surfaces. The diameters of these wet disks may range from less than 1 mm when the precipitation is in the form of mist, to wet spots more than 3 cm during the initial stages of a violent thunderstorm.

There are a number of easy methods of obtaining a record of the size and size distribution of liquid precipitation. (The method for replicating frozen precipitation is described on pp. 288–90.)

One of the simplest and most convenient methods of getting a permanent record is to use Kodak Linograph paper. This is a very slow reacting photosensitive paper, cream-colored prior to exposure to light and grayish afterward. If a raindrop falls on it and it is then exposed for a few seconds to a bright light or for a minute or so to daylight, the wetted portion turns a dark blue color. The diameter of the splash mark on the paper is considerably larger than the drop that made it. This is highly convenient, especially for measuring the smaller droplets. The ratio of splash diameter to drop diameter can be established experimentally by letting a drop of measured volume fall for a distance of 10 meters using polyethylene tubes of various sizes (see p. 300), catching a number of drops, and then determining their weight or volume. Under the same experimental conditions, a strip of sensitized paper is interposed briefly above the water collector so that a number of drops fall on the paper. The splash marks can then be measured and the ratio of splash size and drop diameter established.

The remarkable scientist Wilson Bentley of Vermont devised an elegant method for obtaining the drop size distribution of rain. He took dry flour, sieved it into a shallow pan so that it was fluffy and free of lumps, and then let raindrops fall into it; each raindrop produced a ball of flour dough that was equal to its water volume. He then placed the pan in an oven and baked its contents. After baking, he again sieved the flour and had an excellent visual record of the number and relative sizes of the raindrops. His method can be improved slightly by adding a dry water-soluble dye such as methylene blue to the flour. Thus, any water drop, no matter how small, is colored. This eliminates the possibility of confusing small lumps in the flour with the balls formed by raindrops.

Duncan Blanchard has devised a clever method for measuring the diameter of raindrops. He stretches a patch of women's silk or nylon stocking on a 20-cm diameter embroidery hoop and then dusts the surface of the cloth with powdered sugar. The falling raindrop passes through the open weave of the stocking, leaving a circular hole in the sugar coating. It is essential, however, to work with *used* stockings; the sizing in new ones makes it difficult to coat the surface with a uniform layer of the powdered sugar.

It should be remembered that raindrops larger than about 3 mm

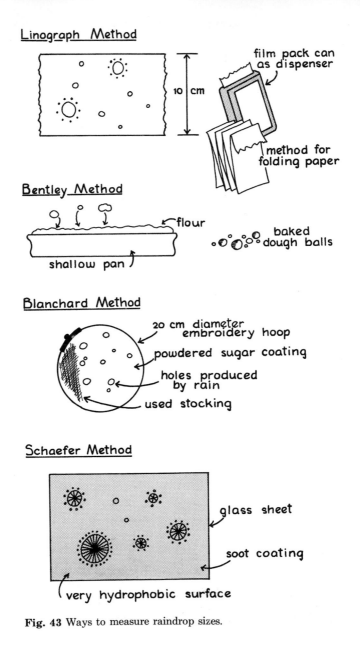

Fig. 43 Ways to measure raindrop sizes.

are likely to leave holes somewhat larger than their spherical diameter — air resistance flattens the larger raindrops when they fall at terminal velocity so that they are convex above but slightly concave below, thus exhibiting a diameter larger than might be expected.

A fourth method for measuring as well as exploring the splash marks of raindrops produces a very complex and beautiful pattern. A sheet of glass is sprayed with silicone oil or rubbed with melted ferric stearate (an orange- or red-colored waxy metallic soap obtainable from a chemical supply house). After it is thoroughly rubbed with a dry, clean towel so that no obvious residue remains, the glass surface becomes extremely hydrophobic: water will run off it as from a duck's back or a sheet of clean polyethylene. When this treated surface is passed about 20 times over a yellow flame from a propane torch, it becomes coated with a very thin yellowish or grayish layer of soot. (In order to obtain a yellow flame, cover some of the air holes temporarily with tape. A candle flame is not effective because it produces too much soot.)

For experimental purposes, a drop of water from a medicine dropper should be made to fall from a height of 2 m onto the glass sheet resting on a table or the floor. If the soot coating has been properly prepared, the splash pattern will be complex and beautiful. In the center of the splash mark will be a highly compact pattern (called the impact spot) which is surrounded by many uniform radiating grooves cut into the soot coating. These patterns provide much food for thought and speculation.

XIII. Electricity in the Atmosphere

Nearly everyone is familiar with the outstanding example of electricity in the atmosphere, lightning. This and smaller electrical discharges in clouds are the primary cause of radio static. There are several simple ways to measure atmospheric electricity.

If a needle is fastened to an insulated wire at the top of a 10-meter pole, electricity will flow from the earth to the atmosphere or vice versa. Under fair-weather skies, little if any current flow can be detected with this device since several thousand volts are needed before an ordinary needle can "go into corona." However, if an air-ionizing source such as a cartridge of radioactive polonium[*] is mounted at the top of the pole, it produces a localized zone of ions that serve as an effective electrical "connection" to the atmosphere. Using this device in conjunction with an ultrasensitive microammeter,[†] it is easy to measure currents as low as 0.01 microampere. This is the type of fair-weather current that produces an

---

[*] Nuclear Products Co., P.O. Box 1178, El Monte, Cal. 91734 (3" Model 3C125 Cartridge, $5.95)
[†] R.C.A. Ultrasensitive Microammeter, Princeton, N.J.

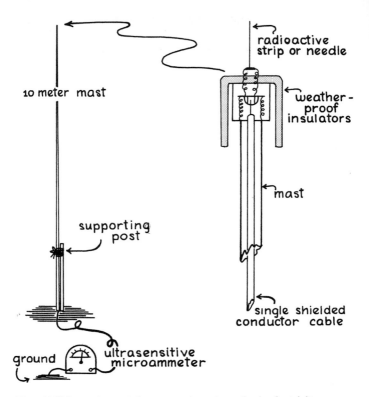

Fig. 44 Telescopic mast for measuring atmospheric electricity.

electrical gradient near the earth of $+60$ to $+100$ volts per meter. Of course, when the weather is severe, microammeter readings should be made from a sheltered place using a long wire.

As the atmosphere becomes cloudy, the corona current increases until values as high as 5–10 microamperes may be observed in the vicinity of a thunderstorm. The electrical gradient under an active thunderstorm often shows values of 5000–10,000 volts (plus or minus) per meter.

There are a number of devices for measuring and studying the electricity of the atmosphere, some of which are highly sophisticated and very expensive. The one described here is a simple, portable, and highly informative device. An ideal pole is a telescopic aluminum radio mast in 4 sections, each about 2 m long. The insulators can be mounted on the top end of the mast with a single-conductor, polystyrene-insulated cable carried down inside the pipe to the sensing instrument.

Various measuring devices, from a strip recorder to the ultrasensitive microammeter, can be used with this detector. With a microammeter it is quite feasible to observe the needle position and record its value about every 2 or 3 seconds. Such data recorded for a period of 15 or 20 minutes can then be plotted on graph paper to provide a useful record of atmospheric electrical conditions. Although somewhat tedious, this method is a very good one since it requires the observer to be present while the cause of the event can be observed, analyzed, and evaluated. Microammeter readings are preferable, but it is also possible to use a simple electronic circuit including a condenser and a tiny neon lamp. Each time the condenser charge reaches a critical value, it discharges through the light, and the number of flashes per unit of time that the current produces indicates the variations that occur.

XIV. The Use of Photography in the Atmosphere

Photography is one of the most useful of all the skills employed in the study of atmospheric phenomena. Its use goes all the way from the photos of large weather systems from our stationary and orbiting earth satellites, to visual records of clouds seen from airplanes and at ground level, and photomicrographs of particles swept from the air on sampling slides.

Suggestions made here will be limited primarily to cloud photography. If these techniques are properly understood and successfully followed, many other uses of the camera will become obvious.

There is almost limitless variety of cameras and film; however, after experimenting and obtaining good photographs, it seems wise to adopt a few standard ones. The opportunities for getting special photographs of rare or unique phenomena in the atmosphere are of such a transient nature that it is important for observers to know their tools and raw materials so well that they can be fairly certain of obtaining a good record.

Excellent photographs of clouds can be obtained in color as well as in black-and-white photographic emulsions. After many years of field experience and experimentation (still underway), I take color photographs with a 35mm camera, using one of 5 lenses — 8, 28, 55, 90, and 360 mm. For most purposes the 55mm lens is adequate. It is good to use an ultraviolet filter at all times.

For black-and-white photography, I use a 10 × 12.5 cm (4 × 5 in.) view camera (a Graflex, for example) that accommodates cut film, film pack, and Polaroid film holders. For cloud photography I use a red Wratten A filter.

After many trials and experiments with cloud photography, I have come to use almost exclusively 35mm Kodak Ektachrome-X daylight color film (ASA 64), 10 × 12.5 cm Kodak Tri-X Panchromatic film (ASA 400), and the Land-Polaroid #55 P/N type film

(ASA 50). A red Wratten A filter works well with the Land-Polaroid #55 film at an ASA setting of 20. Almost any camera that can use the films described will provide good cloud photographs if the sky is fairly clear of pollution haze.

The use of color film is particularly important for recording optical effects such as rainbows, halos, sunstreaks, and other atmospheric phenomena. Using the ultraviolet filter mentioned above seems to provide a faithful rendition of the blue of the sky as well as capturing bluish shadows and the many gradations of cloud color.

One important thing that can be accomplished especially well with the 35mm camera is the preparation of 3-dimensional photo pairs. A single-lens reflex camera is preferable, and the idea is to photograph the same scene from 2 different locations. Unlike the more common stereoscope or antique stereopticon views, 3-dimensional cloud photographs require a base line of 300 m (1000 ft.) or more. This can easily be achieved in an airplane or automobile. The best stereo match is obtained if the reflex viewfinder has either cross hairs or a small circle in the center. This is pointed at a recognizable feature when the first picture is exposed, and at the same feature when the second exposure is made. The time period between the pair should be about 5 seconds in airplanes at cruising altitude; during climb or descent, the pictures should be taken as quickly as possible.

In a few instances when stereos of far-distant clouds are wanted, the time interval can be increased to 10 seconds or even longer. Only experience can provide the best time interval to use. If the photo is quite unusual, 3 or more exposures can be made using 10-second intervals.

With cirrus and other clouds moving rapidly in a jet stream, it is often possible to obtain stereo pairs at a fixed position on the ground. The best results are achieved if the first exposure is made slightly upwind of the cloud stream so that the second photograph is taken slightly downwind.

Stereo pairs will reveal features of clouds that were never suspected. Clouds cannot really be "seen" until they have been viewed in 3 dimensions using either a hand-held slide stereoscope, 2 hand-held viewers, or a pair of slide projectors equipped with polascreen filters, a special viewing screen, and polascreen glasses.

Small droplets or particles of water, ice, dust, or smoke produce a considerable variety of optical phenomena in the atmosphere — scattering, diffracting, or reflecting the light of the sun, moon, stars, and other light sources.

Photographing such effects is not easy, although very satisfying if the effort is successful. It is well to remember that the phenomenon to be photographed is a source of light. Because it is localized and thus does not contribute much additional light for measurement with an exposure meter, there is a tendency on the part of most people to depend on the exposure meter value. It is best to close the aperture from ½–1 full stop more than is indicated. If the effect is unusual, 4 exposures should be made starting with the value indicated by the meter and then closing the aperture ½, 1, and 1½ stops.

Lightning photography is another fascinating activity. While it is possible to obtain lightning photos in the daytime, it is extremely unusual for them to be of much value, either from the aesthetic or scientific standpoint, except for the specialist; much of the intricate pattern of the lightning channel is lost because of the lack of contrast. Lightning photography at night is a different matter. If at all possible, the site for taking the photographs should be completely dark, although distant lights are not troublesome and may even add to the interest of the photo. The camera is best mounted on a tripod, although excellent pictures are often obtained with the camera hand-held or resting on a windowsill.

In preparing to photograph, the pattern of strikes should be noted, as well as the time between successive strikes. The best and most interesting lightning pictures are obtained at distances ranging from a couple hundred meters to about 8 kilometers (5 mi.). If the photographer is close to the storm, he should be under cover and is safest in a car or other metal-framed structure. In a building or on a sheltered porch, there should be no window to cause spurious reflections. The types of film recommended for cloud photography work very well for lightning. Once the average time interval between strikes has been noted (it often is about 10 seconds), the lens should be opened with "time" or "bulb" setting several seconds before the strike is anticipated, and left open until the strike occurs. It is best to limit the exposure to a single observable strike since a second one might obscure some interesting feature of the first. A handy fact about thunder with lightning is that sound travels roughly 1.6 km (1 mi.) in about 5 seconds. The time period between lightning and thunder should be noted: if it is 2 seconds or less, it is wise to seek some type of safe shelter.

It is interesting to experiment with infrared color and other films as time and opportunity permit. Using color film, it is possible to obtain remarkable lightning photographs with a diffraction grating in front of the camera lens.

One final aspect of cloud photography is the preparation of time-lapse movies of clouds. Here are a few basic points for success, based on much experimentation and long experience.

The camera should always be mounted on a tripod fitted with a panoramic head, which should be level. The best time interval for most cloud photos is 1 frame every 2 seconds. Excellent life cycle "sequences" of small clouds can be obtained in 20–30 minutes with an ordinary 16mm or Super 8 movie camera if it has a control that exposes 1 frame at a time. Hand operation can become almost automatic with a little practice. If the clouds are moving rapidly, the initial exposures should be directed toward the upwind direction of cloud movement. When a new orientation of the camera is needed, the panoramic head should be moved not more than 1° at a time, with 1 frame exposure for each change of orientation. Frequent checks on light meter readings should be made, especially if a large storm is in development stages.

Photography is an essential tool for those who want to become seriously involved in atmospheric observation.

XV. Surface Chemistry in the Atmosphere

Surface chemistry, which involves the study of single layers of molecules on liquid or solid surfaces, has many important applications to the atmosphere.

Many airborne particles are surface-active, especially in polluted atmospheres. Such particles will exude molecules when placed on a water surface. In the natural atmosphere, pollen grains — which are all somewhat waxy — are particularly interesting in this respect; when they land on water, they lay down a monomolecular layer of wax that is often a hundred times larger in diameter than the particle producing it.

Some of these properties can be detected and studied with the Langmuir trough. This is a very simple but highly useful device, in essence a shallow tray with straight, level sides whose inside edges are fairly sharp. If the tray is plastic and has these characteristics, no further treatment is necessary. If the tray is metal, the edges must be cleaned and rubbed with paraffin wax.

The important property of a Langmuir trough is that it can be filled with water to a level higher than its edges without overflowing. This feature allows the water surface to be cleaned of surface-active materials.

An ideal size for such a trough is 15 cm wide × 50 cm long × 2.5 cm deep. Two chromium-plated metal strips at least 20 cm long and 0.5 × 0.5 cm in cross section are needed for cleaning the surface of the water in the tray. Suitable strips may be obtained in a variety store in the form of kitchen towel racks.

It is important that the water-filled trough be level. This can be accomplished by resting it on a 1-cm thick sheet of marine plywood as wide as the tray and 5 cm longer. Three or 4 threaded rods 5 cm long serve as legs for this leveler. Projecting 2 cm above its surface, they permit adequate control of the water level.

It is also a good idea to have this assembly resting on a larger trough to take care of spillage; the larger trough can be easily constructed from a wooden or corrugated cardboard box lined with plastic sheeting.

One final consideration is the adequate illumination of the water surface. This can be accomplished as shown in Fig. 45.

To increase the visibility of surface effects, the tray should be black. If the tray is plastic, black paint can be applied to the outside. If the tray is metal, it can be lined with black glass or painted black on the inside. However, in the latter case it is important that the paint used is highly stable so that some of its components do not leach out to contaminate the water with surface-active materials.

Water is added to the trough and leveled so that its surface is at least 1 mm above the hydrophobic edges. The light source must be oriented so that its white surface is reflected on the water. A cleaning bar is placed at one end of the tray and moved at right angles to the length of the tray from one end to within 1 cm of the other. This sweeps any surface films that may be present and crumples them at the far end of the tray. The second bar is applied in a similar manner. The first one is then retrieved, wiped with a clean towel (cloth or paper), and the surface swept again. If any dust particles, powdered talc, or sulfur are on the surface, they will be pushed to the end of the tray by this operation.

Fig. 45 Langmuir Trough.

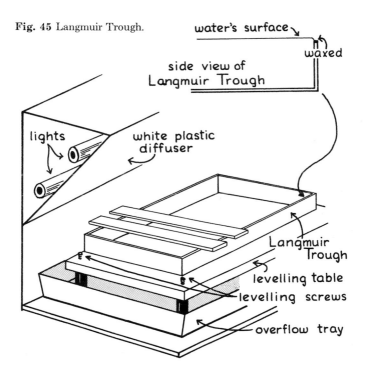

water's surface

waxed

side view of
Langmuir Trough

lights

white plastic
diffuser

Langmuir
Trough

levelling table

levelling screws

overflow tray

In order to make invisible films visible on the water surface, an "indicator oil" is prepared. This has an effect similar to the colored films formed by drops of automobile oil on a wet street after a rain, an interference phenomenon.

Indicator oil is made up of 2 oils. About 25 cubic centimeters of oxidized oil (the oldest motor oil that can be found) is blended with an equal amount of fresh, pure mineral oil without additives which can be obtained from any drugstore under the trade name of Nujol or Petrolatum.

A pointed toothpick or the straightened end of a cleaned paper clip is dipped into the indicator oil and touched briefly to the cleaned water surface. If the oil spreading from the applicator does not show a color like yellow, red, blue, or green, then more of the pure mineral oil is needed.

These adjustments continue until a colored disk appears around the applicator. If it has a bright color, then a small amount of oxidized oil is added and again thoroughly mixed before testing. The desired color is a silver gray. It has a film thickness of about 0.8 micron.

Once this thickness of indicator oil is attained, the Langmuir trough is ready for use. With the old films swept to the end of the tray, a new film of the silvery appearance is applied to the cleaned surface. A bit of experience will indicate the amount of oil needed to cover the surface in one application. If the surface is not quite covered (as indicated by dark areas at the edges or center of the application zone), the last cleaning bar used may be moved against the film edge until the silver color starts to turn yellow.

If at this point a finger is dipped into the oil coating, a black spot will appear surrounding the finger. This is a monomolecular film of surface-active material normally present on hands and many other objects. Although a film such as this tends to be only 0.001 micron thick, it appears to be visible due to the contrast between the silver-colored oil and the monolayer, which looks black because that is the color of the tray.

If this film is swept away and a new film of indicator oil applied and then left alone, within a few minutes it will be pockmarked with a number of black holes. These are monolayers that have been formed by small particles settling onto the water from the air and spreading their individual films over the surface.

Pollen grains, spores, aerosol spray residues, and many other airborne particles can now be tested and compared to each other for surface activity.

Appendixes
Glossary
Bibliography
Index

Types of Frozen Precipitation

CODE	GRAPHIC SYMBOL	TYPICAL FORMS			TYPE
1	⬡				PLATES
2	✳				STELLARS
3	▭				COLUMNS
4	↔				NEEDLES
5	⊕				SPATIAL DENDRITES
6	⊟				CAPPED COLUMNS
7	⋋				IRREGULAR CRYSTALS
8	⧖				GRAUPEL
9	△				SLEET
0	▲				HAIL

APPENDIX 2
International Weather Symbols

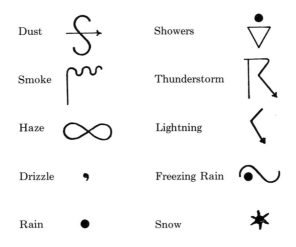

	Dust		Showers
	Smoke		Thunderstorm
	Haze		Lightning
	Drizzle		Freezing Rain
	Rain		Snow

Intensity and Character

Slight	Continuous	Intermittent Moderate at Observation Time	Continuous Moderate at Observation Time	Intermittent Heavy at Observation Time	Continuous Heavy at Observation Time

APPENDIX 3
International Cloud Symbols

Used internationally in weather reporting and plotted on forecasters'
weather charts

Cu, Fc   cumulus, fractocumulus   As   altostratus

Cb   cumulonimbus   Ac   altocumulus

Sc   stratocumulus   Ci   cirrus

St, Fs   stratus, fractostratus   Cs   cirrostratus

Ns   nimbostratus   Cc   cirrocumulus

Low Clouds
(0–3 km)

1. Cu of fair weather, little vertical development and seemingly flattened

5. Sc not formed by spreading out of Cu

2. Cu of considerable development, generally towering, with or without other Cu or Sc bases all at same level

6. St or Fs or both, but no Fs of bad weather

3. Cb with tops lacking clear-cut outlines, but distinctly not cirriform or anvil-shaped; with or without Cu, Sc, or St

7. Fs and/or Fc of bad weather (scud)

4. Sc formed by spreading out of Cu; Cu often present, also

8. Cu and Sc (not formed by spreading out of Cu) with bases at different levels

9. Cb having a clearly fibrous (cirriform) top, often anvil-shaped, with or without Cu, Sc, St, or scud

Middle Clouds
(3-8 km)

1. Thin As (most of cloud layer semi-transparent)

2. Thick As, greater part sufficiently dense to hide sun (or moon), or Ns

3. Thin Ac, mostly semi-transparent; cloud elements not changing much and at a single level

4. Thin Ac in patches; cloud elements continually changing and/or occurring at more than one level

5. Thin Ac in bands or in a layer gradually spreading over sky and usually thickening as a whole

6. Ac formed by the spreading out of Cu

7. Double-layered Ac, or a thick layer of Ac, not increasing; or Ac with As and/or Ns

8. Ac in the form of Cu-shaped tufts or Ac with turrets

9. Ac of a chaotic sky, usually at different levels; patches of dense Ci are usually present, also

High Clouds
(8 km and higher)

1. Filaments of Ci, or "mares' tails," scattered and not increasing

2. Dense Ci in patches or twisted sheaves, usually not increasing, sometimes like remains of Cb; or towers or tufts

3. Dense Ci, often anvil-shaped, derived from or associated with Cb

4. Ci, often hook-shaped, gradually spreading over the sky and usually thickening as a whole

5. Ci and Cs, often in converging bands, or Cs alone; generally overspreading and growing denser

6. Ci and Cs, often in converging bands, or Cs alone; generally overspreading and growing denser

7. Veil of Cs covering the entire sky

8. Cs not increasing and not covering entire sky

9. Cc alone or Cc with some Ci or Cs, but the Cc are the main cirriform cloud

APPENDIX 4
Conversion of Units of Measurement

U.S. to Metric Approximate Conversions

	When You Know	Multiply by	To Find
Length	in	25,000	micron
	in	2.5	cm
	ft	0.3	m
	mi	1.6	km
Area	in^2	6.5	cm^2
	ft^2	0.09	m^2
	mi^2	2.6	km^2
Volume	in^3	16.5	cm^3 (cc)
Mass	oz	28	g
	lb	0.45	kg
Temperature	°F	5/9 after subtracting 32	°C

Metric to U.S. Approximate Conversions

	When You Know	Multiply by	To Find
Length	micron	0.00004	in
	mm	0.04	in
	cm	0.4	in
	m	3.3	ft
	km	0.6	mi
Area	cm^2	0.16	in^2
	km^2	0.4	mi^2
Volume	cm^3 (cc)	0.06	in^3
Mass	kg	2.2	lb
Temperature	°C	9/5, then add 32	°F

Note: in^3 = cubic inch(es); cm^3 = cubic centimeter(s)

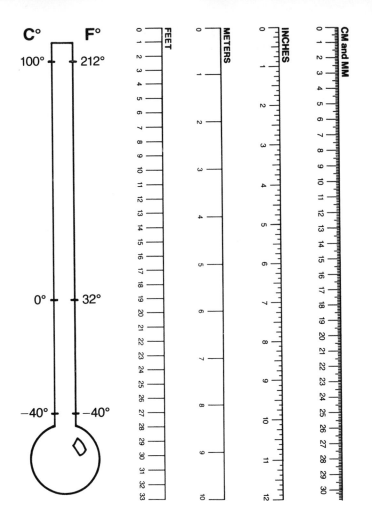

APPENDIX 5
The Electromagnetic Spectrum

μm = micron

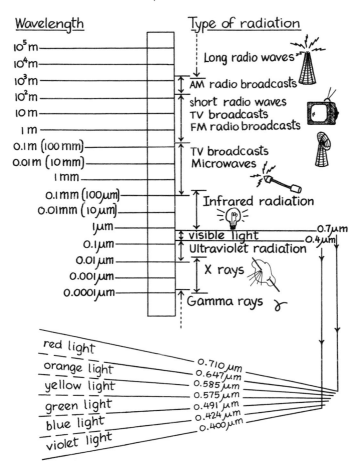

Wavelength	Type of radiation
10^5 m	
10^4 m	Long radio waves
10^3 m	AM radio broadcasts
10^2 m	short radio waves
10 m	TV broadcasts
1 m	FM radio broadcasts
0.1 m (100 mm)	TV broadcasts
0.01 m (10 mm)	Microwaves
1 mm	
0.1 mm (100 μm)	Infrared radiation
0.01 mm (10 μm)	
1 μm	visible light — 0.7 μm
0.1 μm	Ultraviolet radiation — 0.4 μm
0.01 μm	X rays
0.001 μm	
0.0001 μm	Gamma rays γ

red light — 0.710 μm
orange light — 0.647 μm
yellow light — 0.585 μm
— 0.575 μm
green light — 0.491 μm
blue light — 0.424 μm
violet light — 0.400 μm

APPENDIX 6
Seeing

For an object to be seen, light must come from it and experience one or more of five processes. Some major phenomena of the atmosphere, and the processes by which we see them, are summarized below:

Radiation	Reflection	Refraction	Diffraction	Scattering
Sun	Moon	Rainbow	Corona	Sky Color
Lightning	Undersun	Fog Bow	Halo	Cloud Color
Fire	Cloud	Green Flash	Aureole	Crepuscular
Meteorite	Rain, Snow,	Halo	Rainbow	Ray
St. Elmo's	and Hail	Mirage	Mirage	Pollution
Fire	Contrail	Snow Crystal	Ice Cloud	Haze
Aurora	Water Surface	Icicle		Whiteout
	Ice Surface	Parhelia		Blue Haze
	Tornado			Snow Shadow
	Dust Devil			

The first process is *radiation.* If the body is to be seen, the body must be hot enough to emit photons with wavelengths between 0.4 and 0.7 micron which can stimulate the optic nerve receptors of the viewer. Because we live in an electromagnetic environment, all bodies are continuously emitting streams of photons, and by the same token, are being bombarded by photons that have been emitted by other radiant bodies. The energy carried by an individual photon is much too small to be noticed except by the most sensitive instrument. It is only the overall effect that produces noticeable results for the viewer.

The second process is *reflection.* Most objects are seen because of light from some other luminous body—such as the sun or an electric lamp—that is reflected from their surface. Their brightness is determined by the number of photons that bounce off their surfaces and come to the eye of the observer; their color is determined by the proportion of light with different energies and thus different wavelengths.

In the process of *refraction,* light is bent by passage from one transparent medium to another of different density—two different masses of air, for example, or air and a glass lens or prism. This bending results in color effects and image displacements in the atmosphere.

Diffraction is a complex phenomenon whose explanation goes beyond the bounds of this *Field Guide* (but is well described in basic physics texts). Light interacting with an array of small particles, like water droplets, is changed in direction with certain consequent variations in perceived color.

Scattering begins when a molecule immersed in a stream of light absorbs some energy from the stream and becomes "excited." It then sheds the excess energy by becoming a light-emitting body itself. By giving off a stream of photons in all directions the small object scatters light out of and away from the initial light beam in which it is immersed. Significant optical and meteorological effects take place when there are many scatterers present. (This subject is amplified in the section on the *Blue of the Sky,* pp. 155–57).

APPENDIX 7

Sky Colors at Sunrise and Sunset
(for clean air)

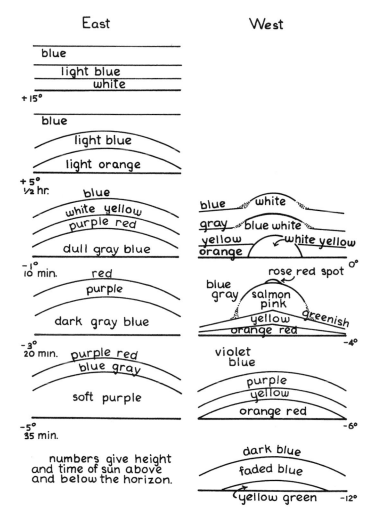

East   West

blue
light blue
white
+ 15°

blue
light blue
light orange
+ 5°
½ hr.

blue   blue   white
white yellow gray blue white
purple red   yellow   white yellow
dull gray blue orange
−1°
10 min.   0°

red   rose red spot
purple   blue
gray   salmon
pink
dark gray blue   yellow   greenish
orange red
−3°   −4°
20 min.

purple red   violet
blue gray   blue
purple
soft purple   yellow
orange red
−5°   −6°
35 min.

numbers give height
and time of sun above   dark blue
and below the horizon.   faded blue
yellow green   −12°

329

Twilight Phenomena

Variations in width of shaded areas indicate intensity of phenomena at different times during twilight.

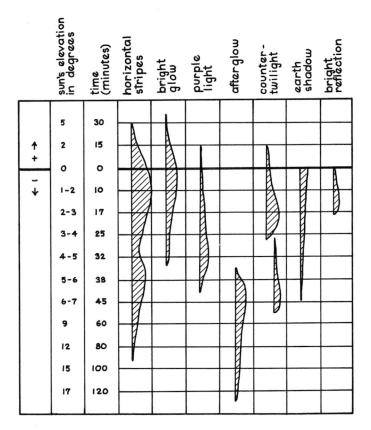

Beaufort Numbers for Wind Force—
Equivalent Wind Speeds and Effects

Beaufort Number	Force	Km/hr
0	Calm	< 1
1	Light air	1–5
2	Light breeze	6–11
3	Gentle breeze	12–19
4	Moderate breeze	20–29
5	Fresh breeze	30–38
6	Strong breeze	39–50
7	Moderate gale	51–61
8	Fresh gale	62–74
9	Strong gale	75–86
10	Whole gale	87–101
11	Storm	102–120
12	Hurricane	>120

Beaufort Number	Mi/hr	Effects
0	< 1	Smoke rises vertically; no perceptible movement of anything.
1	1–3	Smoke drift shows wind direction; barely moves tree leaves.
2	4–7	Wind felt on face; leaves rustle; small twigs move.
3	8–12	Leaves and small twigs in constant motion; blows up dry leaves from ground.
4	13–18	Moves small branches; raises dust and paper and drives them along.
5	19–24	Large branches and small trees in leaf begin to sway; crested wavelets form on inland water.
6	25–31	Large branches in continuous motion.
7	32–38	Whole trees in motion; inconvenience in walking.
8	39–46	Breaks twigs and small branches; difficult to walk.
9	47–54	Loosens bricks on chimneys; blows roofing slates off; litters ground with broken branches.
10	55–63	Trees uprooted; considerable structural damage.
11	64–75	Widespread damage.
12	> 75	Severe and extensive damage.

APPENDIX 10
Wind and Pressure Relationships

"Stand with your back to the wind. High pressure is to your right; low pressure is to your left." This is Buys Ballot's Law (for the northern hemisphere), a convenient handrule relating atmosphere pressure to wind direction.

Buys Ballot's Law is the practical expression of a meteorological equation based on the Geostrophic Wind. The Geostrophic Wind results when there is a balanced relationship between the pressure gradient force and the Coriolis force (see below).

As water at high elevation is acted upon by gravity to run downhill, so air in regions of high pressure is acted upon by gravity to move toward low pressure. The pressure gradient is the maximum rate of change of pressure between two reference points at the same elevation. The larger the pressure gradient, the stronger the wind (see chart).

If the earth did not rotate, pressure differences would be equalized by airflow from the regions of high pressure to regions of lower pressure. Rotation of the earth creates the Coriolis effect, which causes the atmosphere to be deflected. At the equator the Coriolis effect is zero; at the poles it is maximum. Its strength is proportional to the wind speed. It acts at right angles to the wind direction and in the northern hemisphere pulls the moving air to the right, as sensed by an observer with his back to the wind.

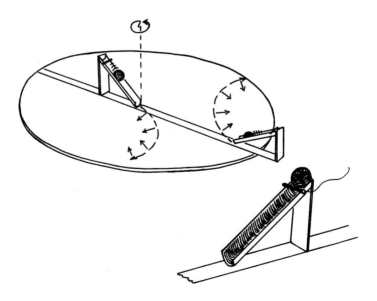

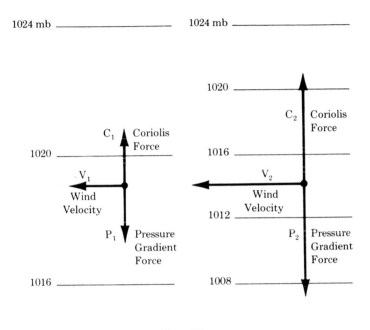

$$V_2 = 2V_1$$
$$C_2 = 2C_1$$
$$P_2 = 2P_1$$

Concentration of Cloud Particles and Precipitation Particles

Particles per cubic meter	1	10^1	10^2	10^3	10^4	10^5	10^6	10^7	10^8	10^9	10^{10}	10^{11}
Particles per liter	10^{-3}	10^{-2}	10^{-1}	1	10^1	10^2	10^3	10^4	10^5	10^6	10^7	10^8
Particles per cubic centimeter	10^{-6}	10^{-5}	10^{-4}	10^{-3}	10^{-2}	10^{-1}	1	10^1	10^2	10^3	10^4	10^5

CLOUD PARTICLES

Cirrus

Ice Cloud

Ocean Fog

Oceanic Clouds

Continental Fog

Stratus

Cumulus

PRECIPITATION PARTICLES

Hail — from Cumulus

Graupel — from Cumulus

Dusty Snow — from Ice Cloud

Misty Rain — from Continental Fog

Drizzle — from Stratus and Fog

Sleet — from Cirrus

Light Snow — from Cirrus

Moderate Snow — from Cumulus

Heavy Snow — from Cumulus

Sea Mist — from Ocean Fog

Heavy Rain — from Cumulus

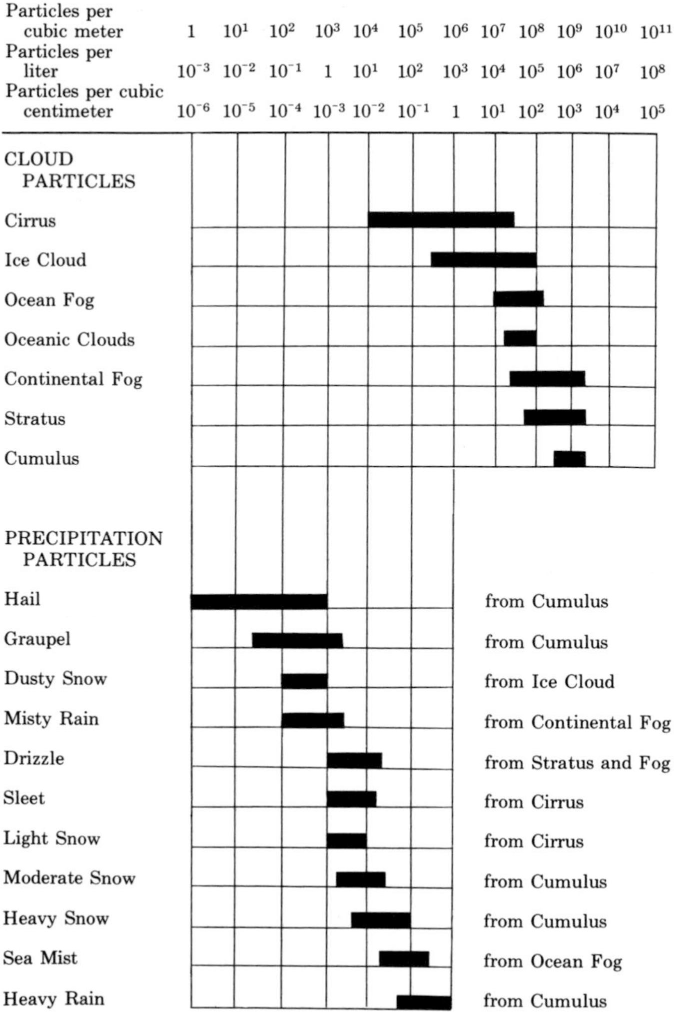

Note: $10^1 = 10$

Principal Features of Halo Phenomena

HALO	refracting angle of min. deviation	orientation of principal crystal axis
22°	60°	random; incident ray 90° to principal axis
46°	90°	random: incident ray 90° to principal axis
PARHELIA	60°	vertical; intensely low when sun is high
CIRCUMSCRIBED TO 22° HALO	60°	horizontal
PARRY ARCS	60°	horizontal
CIRCUM-ZENITHAL ARC	90°	vertical
CIRCUM-HORIZONTAL ARC	90°	vertical
LATERAL TANGENT ARCS OF 22° HALO (LOWITZ ARC)	60°	oscillating
INFRALATERAL TANGENT ARCS OF 46° HALO	90°	horizontal
SUPERLATERAL TANGENT ARCS OF 46° HALO	90°	horizontal

APPENDIX 13

Standard Atmosphere Properties

Altitude (m)	Temperature (°C)	Pressure (mb)	Density (kg/m³)
0	15.0	1,013.2	1.2250
500	11.8	954.6	1.1673
1,000	8.5	898.8	1.1117
1,500	5.2	845.6	1.0581
2,000	2.0	795.0	1.0066
2,500	− 1.2	746.9	0.9569
3,000	− 4.5	701.2	0.9092
3,500	− 7.7	657.8	0.8634
4,000	−11.0	616.6	0.8194
4,500	−14.2	577.5	0.7770
5,000	−17.5	540.5	0.7364
5,500	−20.7	505.4	0.6975
6,000	−24.0	472.2	0.6601
6,500	−27.2	440.8	0.6243
7,000	−30.4	411.0	0.5900
7,500	−33.7	383.0	0.5572
8,000	−36.9	356.5	0.5258
8,500	−40.2	331.5	0.4958
9,000	−43.4	308.0	0.4671
9,500	−46.7	285.8	0.4397
10,000	−49.9	265.0	0.4140
15,000	−56.5	121.1	0.1948
20,000	−56.5	55.3	0.0889
30,000	−46.6	12.0	0.0184
40,000	−22.8	2.9	0.0040
50,000	− 2.5	0.8	0.0010
60,000	−17.4	0.225	0.000306
70,000	−53.4	0.055	0.000088
80,000	−92.5	0.010	0.000020
90,000	−92.5	0.002	0.000003

Adapted from the *U.S. Standard Atmosphere*. 1962. Superintendent of Documents, U.S. Government Printing Office, Washington, D.C.

Record Highs and Lows in the U.S. (through 1976)

State	Record High		Record Low	
	°C	°F	°C	°F
Alabama	45	112	−31	−24
Alaska	38	100	−62	−80
Arizona	53	127	−40	−40
Arkansas	49	120	−35	−29
California	57	134	−43	−45
Colorado	48	118	−51	−60
Connecticut	41	105	−36	−32
Delaware	43	110	−27	−17
Florida	43	109	−19	− 2
Georgia	45	112	−17	− 1
Hawaii	38	100	−10	−14
Idaho	48	118	−51	−60
Illinois	47	117	−37	−35
Indiana	47	116	−37	−35
Iowa	48	118	−44	−47
Kansas	50	121	−40	−40
Kentucky	46	114	−37	−34
Louisiana	46	114	−27	−16
Maine	41	105	−44	−48
Maryland	43	109	−40	−40
Massachusetts	41	106	−37	−34
Michigan	45	112	−46	−51
Minnesota	46	114	−51	−59
Mississippi	46	115	−28	−19
Missouri	48	118	−40	−40
Montana	47	117	−55	−70
Nebraska	48	118	−44	−47
Nevada	50	122	−45	−50
New Hampshire	41	106	−43	−46
New Jersey	43	110	−37	−34
New Mexico	47	116	−46	−50
New York	42	108	−47	−52
North Carolina	43	109	−31	−23
North Dakota	49	121	−51	−60
Ohio	45	113	−40	−39
Oklahoma	49	120	−33	−27
Oregon	48	119	−48	−54
Pennsylvania	44	111	−25	−13
Rhode Island	39	102	−31	−23
South Carolina	44	111	−25	−13
South Dakota	49	120	−50	−58
Tennessee	45	113	−36	−32
Texas	49	120	−31	−23
Utah	47	116	−45	−50
Vermont	41	105	−45	−50
Virginia	44	110	−34	−29
Washington	48	118	−44	−48
West Virginia	45	112	−38	−37
Wisconsin	46	114	−48	−54
Wyoming	46	114	−54	−66

Relative Humidity (percent) with Corresponding Dry Bulb Temperature and Depression of Wet Bulb Thermometer

°F

Dry Bulb	Depression of Wet Bulb Thermometer									
Thermometer	2	4	6	8	10	15	20	25	30	35
20	70	40	12	—	—	—	—	—	—	—
25	74	49	25	1	—	—	—	—	—	—
30	78	56	36	16	—	—	—	—	—	—
35	81	63	45	29	10	—	—	—	—	—
40	83	68	52	37	22	—	—	—	—	—
45	86	71	57	44	31	—	—	—	—	—
50	87	74	61	49	38	10	—	—	—	—
55	88	76	65	54	43	19	—	—	—	—
60	89	78	68	58	48	26	5	—	—	—
65	90	80	70	61	52	31	12	—	—	—
70	90	81	72	64	55	36	19	3	—	—
75	91	82	74	66	58	40	24	9	—	—
80	91	83	75	68	61	44	29	15	3	—
85	92	84	76	69	62	46	32	20	8	—
90	92	85	78	71	65	49	36	24	13	3
95	93	85	79	72	66	51	38	27	17	7
100	93	86	80	73	68	54	41	30	21	12

°C

Dry Bulb	Depression of Wet Bulb Thermometer					
Thermometer	1	2	3	5	10	15
− 10	69	39	10	—	—	—
− 5	77	54	32	—	—	—
0	82	65	47	15	—	—
5	86	71	58	32	—	—
10	88	76	65	44	—	—
15	90	80	70	52	12	—
20	91	82	74	58	24	—
25	92	84	77	63	32	7
30	93	86	79	67	39	16
35	93	87	81	69	44	23
40	94	88	82	72	48	29
45	94	89	83	73	51	33
50	95	89	84	75	54	37

Condensed from Bulletin No. 1071, U.S. Weather Bureau. Superintendent of Documents, U.S. Government Printing Office, Washington, D.C.

Water Vapor Pressure and Density
in Saturated Air, as a Function of Temperature

Pressure and density, measured under conditions of saturation, vary with the temperature of the air. Note the small but significant difference in values of pressure measured over water versus ice.

Temperature		Pressure (cm of mercury)		Density (g/m^3)
°C	°F	over water	over ice	
− 10	14	0.21	0.19	2.2
− 5	23	0.32	0.30	3.2
0	32	0.46	0.46	4.8
5	44	0.65	—	6.8
10	50	0.92	—	9.3
15	59	1.28	—	12.7
20	68	1.75	—	17.1
25	77	2.38	—	22.8
30	86	3.18	—	30.0
40	104	5.33	—	50.9
50	122	9.25	—	83.0
60	140	14.93	—	130.7
70	158	23.37	—	199.5
80	176	35.51	—	295.9
90	194	52.58	—	428.4
95	203	63.39	—	511.1
100	212	76.00	—	606.2
105	221	90.64	—	715.4
110	230	107.46	—	840.1

Safety and Comfort—Wind Chill and Hypothermia

The presence of wind makes it seem colder than is indicated by the reading of the thermometer. This is a pleasing effect when temperatures are high. When temperatures are low, wind chill becomes a threat to human comfort and may cause death. Clearly, how it affects you depends on other variables such as type of clothing worn, amount of exposed flesh, amount of solar radiation, and your general physical condition at the time.

The wind-chill concept, first applied by Paul Siple in 1939, measures the relative cooling power, or rate of heat removal, from the human body at various combinations of wind speed and temperature. Heat is removed from the skin by processes of evaporation, convection, radiation, and conduction. In the comfort zone there is a nice balance between the production rate of heat by the body and its removal rate from the skin. When the natural cooling rate is less than the rate of body heat production, there is discomfort. Every person who has fanned his or her face to make a hot and humid environment more bearable has attempted to increase the rate of evaporative cooling.

When the wind is strong and the temperature is low, the heat removal rate becomes greater than the body's heat-generating capacity. The excessive lowering of body temperature from wind chill or for any other reason is called hypothermia, and causes many winter deaths. It works in this way: When your body loses heat faster than it is being produced, you exercise to stay warm, either voluntarily, or involuntarily by shivering. The body also makes other complex adjustments to maintain the normal temperature of the vital organs. Both responses drain energy reserves. When energy reserves are exhausted, the lowered temperature reaches the brain, depriving it of judgment and reasoning power. Control of your hands and feet is lost. As the body's internal temperature decreases to a critical level, lassitude and stupor come on, followed by collapse and in some cases death.

To avoid hypothermia: (1) be aware of and understand wind chill; (2) stay dry if possible, because wet clothes lose about 90% of their insulating value (*note:* of all fabrics, wool is the best under these conditions); (3) if you cannot stay warm and prevent shivering, seek shelter from wind and precipitation; (4) if possible, drink a hot beverage (not alcohol); (5) get into dry clothes if available; (6) conserve and restore the body's energy with food and rest.

Wind-chill Temperatures Compared to Ambient Temperatures

AMBIENT AIR TEMPERATURE (upper left is Celsius; lower right is Fahrenheit)

EQUIVALENT CHILL TEMPERATURE

Wind Speed km/hr	mi/hr	5/40	2/35	-1/30	-3/25	-6/20	-9/15	-12/10	-15/5	-18/0	-21/-5	-23/-10	-26/-15	-29/-20	-32/-25	-34/-30	-37/-35	-40/-40
CALM																		
8	5	2/35	-1/30	-3/25	-6/20	-9/15	-12/10	-15/5	-18/0	-21/-5	-23/-10	-26/-15	-29/-20	-32/-25	-34/-30	-38/-35	-40/-40	-43/-45
17	10	-1/30	-6/20	-9/15	-12/10	-15/5	-18/0	-23/-10	-26/-15	-29/-20	-32/-25	-38/-35	-40/-40	-43/-45	-46/-50	-51/-60	-54/-65	-56/-70
25	15	-3/25	-5/15	-12/10	-18/0	-21/-5	-23/-10	-25/-20	-32/-25	-34/-30	-40/-40	-43/-45	-46/-50	-51/-60	-54/-65	-56/-70	-62/-80	-65/-85
33	20	-6/20	-12/10	-15/5	-18/0	-23/-10	-26/-15	-32/-25	-34/-30	-38/-35	-43/-45	-46/-50	-51/-60	-54/-65	-59/-75	-62/-80	-65/-85	-71/-95
42	25	-9/15	-12/10	-18/0	-21/-5	-26/-15	-29/-20	-34/-30	-38/-35	-43/-45	-46/-55	-51/-60	-54/-65	-59/-75	-62/-80	-68/-90	-71/-95	-76/-105
50	30	-12/10	-15/5	-18/0	-23/-10	-29/-20	-32/-25	-34/-30	-40/-40	-46/-50	-48/-55	-54/-65	-56/-70	-62/-80	-65/-85	-71/-95	-73/-100	-79/-110
58	35	-12/10	-15/5	-21/-5	-23/-10	-29/-20	-34/-30	-38/-35	-40/-40	-46/-50	-51/-60	-54/-65	-56/-75	-62/-80	-68/-90	-73/-100	-76/-105	-62/-115
67	40	-12/10	-18/0	-21/-5	-26/-15	-29/-20	-34/-30	-38/-35	-43/-45	-48/-55	-51/-60	-56/-70	-59/-75	-65/-85	-71/-95	-73/-100	-79/-110	-62/-115

WINDS ABOVE 40 mi/hr HAVE LITTLE ADDITIONAL EFFECT

LITTLE DANGER

INCREASING DANGER (Flesh may freeze within 1 minute)

GREAT DANGER (Flesh may freeze within 30 seconds)

DANGER OF FREEZING EXPOSED FLESH FOR PROPERLY CLOTHED PERSONS

341

Safety and Comfort—Temperature and Humidity

Warm, dry air is more comfortable than warm, very moist air unless it is *too* dry. An extremely humid, warm day makes one think the temperature is higher than it really is; an extremely dry, hot day may seem to be cooler than the thermometer registers. The "humiture" is a figure in degrees at which air of a given temperature and moisture content "feels like" an equivalent dry temperature. The concept of humiture is based on sound biometeorological principles.

"Humits" are the units that are added to the temperature to obtain humiture. To get humits, subtract 10 from the number of millibars of vapor pressure in the air (*note:* there are no negative humits). For example, a temperature of 90°F and a vapor pressure of 8 mb yields zero humits and a humiture of 90, while a temperature of 90°F and a vapor pressure of 22 mb yields a humiture of 102.

Figuring humiture is simplified by the fact that the number of humits does not change during the day if the air mass remains the same; so constant humits may be added to the ambient temperature at any time of day.

Humits can also be obtained directly from the dew point temperature, as by the conversion table.

Dew point (°F)	50	55	60	65	70	75	80	85
Humits	2	5	8	11	15	19	25	31

Example:

Air Temp.   95
Dew Pt. Temp. 85
Humits   31
Humiture $= 95 + 31 = 126$,
   severe discomfiture

A humiture of 70–85 is considered to bracket the comfort zone for most people—at about 85, discomfort starts to set in for some. Nearly everybody is uncomfortable at 100. At 115 or higher, discomfort is at the point where activity must be restricted.

APPENDIX 19
Tornado Safety Rules

1. During tornado season, have your game plan worked out in advance; work out a tornado drill in your family, school, or business.
2. When a tornado alert is sounded, get set. Keep a battery-powered radio turned on to your local station, or to one of the civil defense stations, 640 kHz or 1240 kHz.
3. Watch any thunderstorm clouds for signs of a funnel; know who the local authorities are and how to reach them; notify the local authorities immediately if you see a funnel.
4. If you have no "cyclone cellar," the next best place for shelter is the southwest corner of the basement. Tornadoes usually move from the southwest and if a house is ripped down, the pieces are most likely to fall into the northeast part of the basement. Get under a heavy table or mattress if possible.
5. Open windows and doors on the lee side (usually northeast) to allow internal adjustments to rapid external pressure changes.
6. Funnels may not be visible at night or during heavy rain. If a loud roar is heard, it means that the funnel is very close; take protective measures at once. If this happens while you are sleeping, immediately crawl under the bed; it will provide protection from flying glass and other dangerous objects.
7. If the tornado approaches a school, hospital, or other densely populated building, the situation is particularly dangerous. If the tornado is far enough away, evacuation to a nearby ravine or deep ditch provides more safety than staying inside. Otherwise, occupants should sit close to an interior wall.
8. Tornadoes usually move at speeds of 32–48 km/hr (20–30 mi/hr); they can be outrun by persons in cars. If this is not possible because of road direction or congestion, get out of the car, find a ditch, and lie low while the funnel passes.
9. If a funnel passes directly overhead, try to keep presence of mind and remember what you see; write it down and send it to either of the authors. Take a picture, if possible, without risking life or limb.

Lightning Safety Rules

If indoors
1. Stay indoors and away from open doors and windows; disconnect plug-in electrical appliances if they are not needed; avoid metal objects like stoves, pipes, and sinks.
2. Don't use plug-in electrical equipment during the lightning storm.
3. Don't use the telephone during the storm unless it is necessary to call for help. Lightning may enter the house via the telephone lines.
4. Don't go outside to take laundry off the clothesline.

If outdoors
1. Cease work and get away from metallic objects. Tractors and other farm equipment in metallic contact with the ground are often struck by lightning.
2. Get in an automobile, or stay in your car if traveling. Automobiles offer excellent lightning protection.
3. Seek shelter in buildings. If no buildings are available, your best protection is in a cave, ditch, or canyon, or under head-high clumps of trees or bushes in open forest glades.
4. When there is no shelter, avoid the highest object in the area. If only isolated trees are nearby, your best protection is to crouch in the open, keeping at least twice as far away from isolated trees as the trees are high.
5. If you are hiking and have a backpack, put backpack on the ground and kneel on it as a makeshift insulator.
6. If your hair stands on end, your skin tingles, or, at night, a bluish glow (St. Elmo's Fire) appears on metal objects, lightning may be about to strike. Drop to the ground immediately.

Note: Persons struck by lightning receive a shock and may be burned but can be handled safely. A person struck by lightning can often be revived by prompt mouth-to-mouth resuscitation, cardiac massage, and prolonged artificial respiration.

Glossary

Adiabatic process: A thermodynamic process in which no heat passes into or out of the working substance (such as air); all expansions use internal sources of energy for the work required.

Air mass: A large body of air (hundreds to thousands of square kilometers) that possesses about the same temperature and moisture content level by level.

Aerosol: A group of small particles, usually droplets, "floating" in the air.

Anticyclone: A large-scale circulation of winds, clockwise in the Northern Hemisphere, around a central region of relatively high atmospheric pressure.

Bénard cells: Cellular convection cells that occur in unstable fluids, initially at rest.

Bimodal distribution: A size or mass spectrum of atmospheric particles that has 2 peaks of concentration.

Black stratus: Mist particles, produced in trade-wind clouds, that are too small to fall rapidly and that scatter light over a very narrow angle.

Brownian coagulation: The process by which a particle grows by the rapid condensation of gases and fine particles.

Cloud street: A straight or curved line of convective clouds such as cumulus.

Cold front: The boundary between a mass of warm air and a mass of cold air that underruns and displaces it.

Collimated: Made parallel, as with parallel light beams.

Cloud condensation nuclei: Small particles of dust, smoke, or salt that serve as centers for condensation.

Coalescence: Fusing together; the process by which a large raindrop grows larger by absorbing smaller drops in its fall path.

Conduction: Transfer of heat from one place to another by molecular agitation and without movement of the medium.

Contrail: A streamer cloud of water drops or ice crystals that forms in the wake of an aircraft, resulting from condensation of moisture added to the air by combustion of fuel.

Convection: Transfer of heat by physical movement of the heated medium from one place to another.

Cyclone: A large-scale system of winds blowing around a center of low pressure; motion is counterclockwise in the Northern Hemisphere, clockwise in the Southern Hemisphere.

Cup scroll: An ice-crystal formation that resembles a cup or spiral.

Counter-twilight: A colored border of 6°–12° in height above the eastern horizon a half hour before sunset showing transitions to orange, yellow, green, and blue.

Dendritic crystals: Crystals that grow in a branching form resembling trees.

Dew point: Temperature at which condensation begins on a surface; relative humidity is 100%.

Dust devil: A small whirlwind common in dry regions and made visible by the dust and debris picked up from the ground.

Earth shadow: The curved red-to-blue shadow seen in the east shortly after sunset.

E.R.T.S.: Earth Resources Technology Satellite.

E.S.S.A.: Environmental Sciences and Services Authority — the predecessor of NOAA. Satellites are designated by numbers that follow these initials.

Foam volcanoes: Ice formations caused by the freezing of foam that emerges from ice-covered streams or lakes.

Frazil ice: Disk-shaped plates of ice that form in streams.

Gas-to-particle reaction: The formation of particles by the rapid condensation of a gas.

Giant salt nuclei: Salt crystals in the atmosphere originating from the evaporation of salt spray.

Glory: A colored halo opposite the sun.

Graupel: A type of precipitation consisting of frozen cloud droplets that have clumped and formed pellets.

Halo: A colored ring, appearing at 22° or 46° around the sun or moon, caused by refraction of light through ice crystals.

Heat sink: That portion of a thermodynamic system that absorbs unused heat.

Hurricane: A violent tropical cyclonic storm of the western Atlantic.

Hydrometeor: Any form of precipitation (snow, rain, hail, etc.)

Hydromite: An ice cylinder, like a cave stalagmite, formed by falling drops of water.

Hygroscopic: Absorbing or attracting moisture, as with salt crystals.

Instability: A layering of the atmosphere that tends toward vertical overturning.

Inversion: Reversal of the normal temperature change with altitude, where temperature rises rather than falls as altitude increases.

Ion: An atom or molecule that has an electric charge.

Jet stream: A fast-moving ribbon of air that occurs near the top of the troposphere; width is 25–100 km, thickness 1–3 km, and velocity 90–300 km per hour.

Lapse rate: The rate of temperature change with increasing altitude.

Laminar flow: Smooth, nonturbulent flow of a gas or liquid.

Latent heat: Internal energy in the form of heat *released* when a change of state occurs from gas to liquid or liquid to solid. Heat is *required* to bring about a change of state from solid to liquid or liquid to gas.

Limb: The edge of a celestial body as it appears to an observer; the edge of the sun's disk, for example.

Natural atmosphere: The atmosphere when it is not affected by human activity.

NOAA: National Oceanographic and Atmospheric Administration. Suffix numerals designate satellite numbers.

Nucleation: The beginning of the process of droplet or ice-crystal formation; homogeneous nucleation occurs when there is no foreign particle to act as a nucleus.

Occluded front: A complex front that forms when a cold front overtakes a warm front and cuts off warm-sector air from the surface.

Order of magnitude: Ten times; thus, an increase of 2 orders of magnitude is 10×10, or 100 times.

Orographic: Pertaining to mountain ranges; thus, a mountain or hill that causes air to rise over it is an orographic barrier.

Overseeding: Presence in the atmosphere of such a high concentration of cloud condensation nuclei or ice nuclei that all available moisture is used up before drops or crystals grow large enough to fall as precipitation.

Ozone: Gas composed of O_3 molecules (3 atoms of oxygen) found in the atmosphere in minute quantities; most of it originates in the stratosphere, some in thunderstorms.

Parhelic circle: A halo around the sun. See **Halo.**

Parhelion: A bright spot on the 22° and 46° haloes, horizontal with the sun.

Pileus: A cloud formed in saturated air above a convective tower and separated from it.

Polarized light: Light that vibrates only in one plane.

Polygonal array: A pattern that develops in a close-packed group of circular units.

Pressure gradient: The rate of decrease in atmospheric pressure per unit of horizontal distance; used to indicate relative pressures on the earth's surface, and measured in the direction of most rapid decrease.

Relative humidity: The amount of water in the air, expressed as a percentage of the maximum amount that the air could hold at a given temperature.

Sastrugi: A scalloped ridge of compacted, wind-driven snow that moves slowly, like a sand dune.

Saturation: A condition in the atmosphere corresponding to 100% relative humidity; *supersaturation* is a condition corresponding to greater than 100% relative humidity.

Saturation vapor pressure: Pressure exerted by molecules of vapor (water) when air is saturated and in equilibrium with respect to water or ice. Saturated vapor pressure is slightly, but significantly, greater when measured over water than over ice.

Schaefer point: The temperature of $-40°C$ ($-40°F$) at which ice crystals form spontaneously without the need of nuclei.

Squall line: A line along which converging air currents bring about cumulonimbus activity, with resulting heavy, squally precipitation.

Standing wave cloud: A cloud formed at the crest of a wave in the airstream caused by a mountain peak or ridge. Wind blows through the cloud while the cloud remains stationary.

Stable aerosol: A group of fine particles "floating" in the air and in a state of equilibrium with the environment.

Stratosphere: The region of the upper atmosphere extending upward from the tropopause characterized by little vertical change of temperature and very little moisture. Height of base varies from 6 km over the poles to 18 km above the equator.

Sublimation: The conversion of ice directly from solid to vapor, or from vapor to solid. (For the latter process, this term is being supplanted by the term *ice condensation*.)

Submicroscopic: Not resolvable in a microscope.

Supercooled cloud: Cloud in which the water droplets, though at temperatures below freezing, still remain in the liquid state.

Supernumerary bows: Additional bows inside the primary rainbow caused by diffraction effects in drops of different sizes.

Terminal velocity: The maximum velocity at which a falling body moves through a medium, such as air. Pull of gravity balanced by friction resisting force.

Tornado: A localized, violently destructive circular windstorm characterized by a long, funnel-shaped cloud.

Tropopause: The boundary between the troposphere and the stratosphere.

Troposphere: The inner layer of the atmosphere, varying in altitude between 9.5 km and 19 km, within which there is a steady decrease of temperature as altitude increases; all weather occurs in this layer.

Turbidity: "Muddiness" in the atmosphere, caused by large amounts of aerosol in suspension.

Typhoon: A tropical cyclone of the western Pacific Ocean and the China Sea.

Undersun: A mirrorlike image of the sun, reflected by high concentrations of floating ice crystals; appears at the same angle below the horizon as the sun is above it.

Varve: Layers of sediment. Coarse and fine layers occur in lakes, for example, when ice covers the water and allows the fine particles to settle.

Vertical moisture profile: Distribution of moisture at different altitudes above a given area at a certain time.

Virga: Falling precipitation, usually ice, that evaporates before reaching the ground.

Warm front: A weather front along which an advancing mass of warm air rises over a retreating mass of cold air.

Waterspout: A tornado over a body of water, with characteristic pendant funnel hanging from a cumulonimbus cloud.

Wilson-type cloud chamber: An experimental chamber in which supersaturated vapor is forced to condense on nuclei.

Wind shear: A change in wind velocity as altitude changes.

Bibliography

Chapter 1

Hogan, A., et al. 1967. Aitken Nuclei Observations over the North Atlantic Ocean. *Jour. of Applied Met.* 6:726.

Junge, C. 1963. *Air Chemistry and Radioactivity.* New York: Academic Press.

Schaefer, V.J. 1976. The Air Quality Patterns of Aerosols on the Global Scale. Part 1 and Part 2. *ASRC-SUNY* Publ. 406. Albany, N.Y.

Sears, R. 1974. *The World's Weather and Climates.* New York: Crown Publishers, Bounty Books.

Chapter 2

Ludlam, F.H., and R.S. Scorer. 1958. *Cloud Study: A Pictorial Guide.* New York: Macmillan.

Schaefer, V.J. 1970. Time Lapse Photograph of Clouds. *Applied Optics* 9:1817.

Sloane, E. 1952. *Eric Sloane's Weather Book.* Boston: Little, Brown & Co.

Sloane, E. 1955. *Eric Sloane's Almanac and Weather Forecaster.* Boston: Little, Brown & Co.

Sloane, E. 1961. *Look at the Sky.* New York: Duell, Sloan and Pearce.

World Meteorological Organization. *International Cloud Atlas.* Vols. I and II. Geneva, Switzerland.

Chapter 3

Reiter, E.R. 1969. *Atmospheric Transport Processes.* U.S. A.E.C. Div. Tech. Information, Oak Ridge, Tenn.

Schaefer, V.J. 1953. Cloud Forms of the Jet Stream. *Tellus* 5:27.

Schaefer, V.J. 1957. The Relationship of Jet Streams to Forest Wildfires. *Jour. of Forestry* 55:419.

Chapter 4

Humphrey, W.J. 1920. *Physics of the Air.* New York: McGraw-Hill.

Minnaert, M. 1954. *The Nature of Light and Colour in the Open Air.* New York: Dover Publications.

O'Connell, D.J.K. 1958. *The Green Flash and Other Low Sun Phenomena.* Vatican Observatory. New York: Interscience Publishers, Inc.

Scorer, R.S. 1972. *Clouds of the World.* Harrisburg, Pa.: Stackpole Books.

Chapter 5

Kraght, P.E. 1943. *Meteorology for Ship and Aircraft Operation*. New York: Cornell Maritime Press.

Wood, Elizabeth A. 1968. *Science for the Airplane Passenger*. Boston: Houghton Mifflin Co.

Chapter 6

Flora, S.E. 1953. *Tornadoes of the United States*. Norman, Okla.: University of Oklahoma Press.

Gokhale, N. 1976. *Hailstorms and Hailstone Growth*. Albany, N.Y.: State University of New York Press.

Sloane, E. 1960. *The Book of Storms*. New York: Duell, Sloan and Pearce.

Schaefer, V.J. 1960. Hailstorms and Hailstones of the Western Great Plains. *Nubila* 3:18.

Chapter 7

Blanchard, D. 1967. *From Raindrops to Volcanos*. Garden City, N.Y: Doubleday-Anchor, Science Study Series.

Schaefer, V.J. 1966. The Detection of Surface Active Molecules of Airborne Particulates. *Jour. Recherches Atmospheriques* 141.

Schaefer, V.J. 1964. The Preparation of Permanent Replicas of Snow, Frost and Ice. *Weatherwise* 17:1.

Schaefer, V.J. 1978. The Presence of Ozone, Nitric Acid, Nitrogen Dioxide and Ammonia in the Atmosphere. *ASRC-SUNY* Publ. 682. Albany, N.Y.

Chapter 8

Langmuir, I. 1948. The Production of Rain by a Chain Reaction in Cumulus Clouds Above Freezing. *Jour. of Meteorology* 5:175.

President's Advisory Committee on Weather Control. 1957. Final Report. Vols. I and II. Supt. of Documents, U.S. Govt. Printing Office, Washington, D.C.

Schaefer, V.J. 1946. The Production of Ice Crystals in a Cloud of Super-cooled Water Droplets. *Science* 104:457.

Schaefer, V.J. 1960. Experimental Meteorology. *Jour. Applied Math & Physics* (*ZAMP*) 1:153, 217.

Schaefer, V.J. 1960. Serendipity and the Development of Experimental Meteorology. *Proc. Amer. Soc. Civil Engineers*.

Schaefer, V.J. 1969. The Inadvertent Modification of the Atmosphere by Air Pollution. *Bull. Amer. Met. Soc.* 50:199.

Vonnegut, B. 1947. The Nucleation of Ice Formation by Silver Iodide. *Jour. Applied Physics* 18:593.

Chapter 9

Humphrey, W.J., and W. Bentley. 1931. *Snow Crystals*. New York: Dover Publications.

Nakaya, U. 1954. *Snow Crystals, Natural and Artificial.* Cambridge, Mass.: Harvard University Press.

Schaefer, V.J. 1961–1971. *Yellowstone Field Research Expeditions.* Albany, N.Y.: Atmospheric Sciences Research Center.

Schaefer, V.J. 1950. The Formation of Frazil and Anchor Ice in Cold Water. *Trans. Amer. Geophys. Union* 31:885.

Schaefer, V.J. 1958. *The Illustrated Library of the Natural Sciences.* Vols. II and IV. New York: Simon and Schuster.

Chapter 10

Langmuir, I., and V.J. Schaefer. 1937. The Effect of Dissolved Salts on Insoluble Monolayers. *Jour. Amer. Chem. Soc.* 59:2400.

Schaefer, V.J. 1974. Scientific Adventures. *ASRC-SUNY* Publ. 342. Albany, N.Y.

Schaefer, V.J. 1974. Simple Experiments in Atmospheric Physics. *Weatherwise.* 1955–1957; reprinted *ASRC-SUNY* Publ. 268. Albany, N.Y.

Schaefer, V.J. 1956. Atmospheric Electrical Measurements in the Hawaiian Islands. *Geophysica Pura Applicata* 34:3.

Schaefer, V.J. 1948. The Production of Clouds Containing Supercooled Water Droplets or Ice Crystals Under Laboratory Conditions. *Bull. Amer. Met. Soc.* 29:175.

Schaefer, V.J. 1959. Simple Laboratory Apparatus for the Study of Clouds. *Cumulus Dynamics.* London: Pergamon Press.

General

Butcher, S.S., and R.J. Charlson. 1972. *An Introduction to Air Chemistry.* New York: Academic Press.

Craig, R.A. 1965. *The Upper Atmosphere Meteorology and Physics.* New York: Academic Press.

Day, J.A., and G.L. Sternes. 1970. *Climate and Weather.* Reading, Mass.: Addison-Wesley.

Flohn, H. 1968. *Climate and Weather.* New York: McGraw-Hill.

Hidy, G.M. 1972. *Aerosols and Aerosol Chemistry.* New York: Academic Press.

Humphreys, W.J. 1942. *Ways of the Weather.* Lancaster, Pa.: Jaques Cattell Press.

Huschke, R., ed. 1959. *Meteorological Glossary.* Boston, Mass.: American Meteorological Society.

Malone, T.M., ed. 1951. *Compendium of Meteorology.* Boston, Mass.: American Meteorological Society.

McHarg, I.L. 1969. *Design with Nature.* Garden City, N.Y.: Natural History Press.

Riehl, H. 1972. *Introduction to the Atmosphere.* New York: McGraw-Hill.

U.S. Standard Atmosphere ICAO Atmosphere to 20 km. Dec. 1962. Supt. of Documents, U.S. Govt. Printing Office.

Index

References to illustrations are printed in **boldface:** for line drawings and black-and-white plates, the page number is given; for color plates, the plate number, preceded by **C.Pl.,** is given.

The WORLD of CLOUDS

OF WHICH THERE ARE
3 ANATOMICAL SHAPES

mare's tail CIRRUS (ci)

C...

AVERAGE TOPMOST HEIGHT OF ALL CLOUDS (LOWEST at POLES, HIGHEST at EQUATOR)

a THUNDERHEAD *may rise to* 50,000'

"ANVIL" TOP SPREADS IN STORM DIRECTION

webby CIRROSTR...

CIRRUS DENSUS (FALSE CIRRUS)

HOW THE SU...
SHINING THR...

UPDRAFT SENDS RAIN INTO
FREEZING AREA TO FALL AS
HAIL

SH...

LIGHTNING

lump-pattern ALTOCUMULUS (ac)

CUMULONIMBUS
(*thunderhead*)

THE HIGHER, THE
DRIER THE WEATHER

LINE SQUALL

fair weather CUMULUS (cu)

a SHORT, SEVERE SHOWER

WIND SHIFT

FIRST DROPS
ARE LARGE

ADVECTION FOG

WET 4...

ATHERMURAL BY *Eric Sloane* ©19..

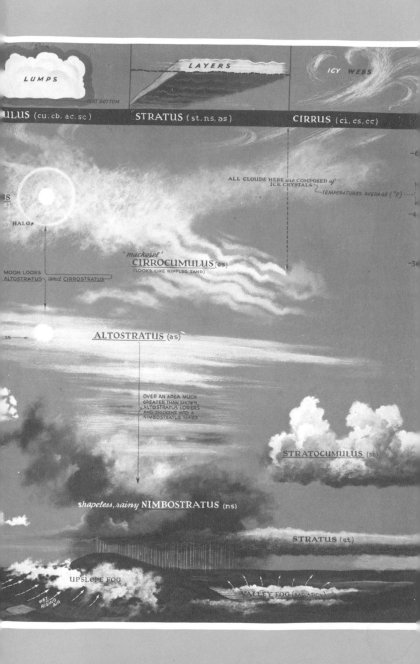

LUMPS

LAYERS

ICY WEBS

— FLAT BOTTOM

...ULUS (cu, cb, ac, sc)   STRATUS (st, ns, as)   CIRRUS (ci, cs, cc)

—6

ALL CLOUDS HERE *are* COMPOSED *of* ICE CRYSTALS
—TEMPERATURES AVERAGE (°F)---

—4

...S

HALO

"mackerel" CIRROCUMULUS (cc)
(LOOKS LIKE RIPPLED SAND)

—3

MOON LOOKS
ALTOSTRATUS *and* CIRROSTRATUS

...SS →   ALTOSTRATUS (as)

OVER AN AREA MUCH
GREATER THAN SHOWN,
ALTOSTRATUS LOWERS
AND THICKENS INTO A
NIMBOSTRATUS MASS

STRATOCUMULUS (sc)

shapeless, rainy NIMBOSTRATUS (ns)

STRATUS (st)

UPSLOPE FOG

VALLEY FOG (RADIATION)

WET
RISING
AIR